Mathematics Puzzles

ISBN 1-58037-148-5

Printing No. CD-1369

Mark Twain Media, Inc., Publishers
Distributed by Carson-Dellosa Publishing Company, Inc.

Table of Contents

Introduction

This mathematics puzzle book presents fundamental math topics in a fun and educational way. The puzzle topics include place value, addition, subtraction, multiplication, division, fractions, decimals, percentages, geometry, algebra, and metrics. Each topic presents information pertinent for carrying out mathematical operations and is followed by a quiz. Each topic also has several puzzles—crosswords, word searches, hidden numbers and messages, and more—for students to complete that will reinforce their knowledge.

Whether used as additional information while studying a certain mathematical topic or as stand-alone activities, students will enjoy learning about math while completing these fun puzzle units as well as developing and strengthening their math skills.

Name ______________________________ Date ______________

Place Value

Place value is the value of a digit in a specific position in a number. Each digit in a number has a specific value depending on what position it holds in the number. For example, each digit in the following number is a 3, but each 3 has a different value depending on its position.

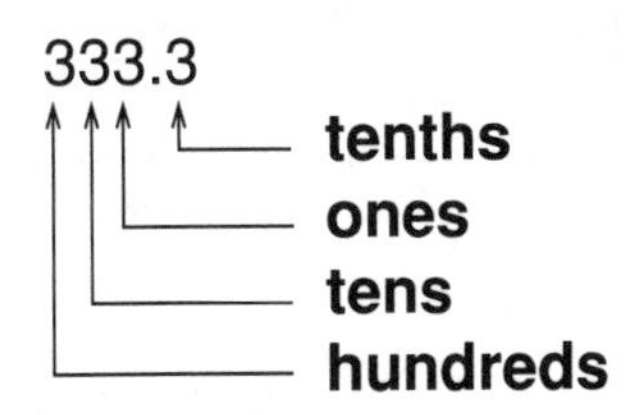

The first 3 equals 300. The second 3 equals 30. The third 3 equals three. The fourth 3 equals 0.3.

When pronouncing numbers, it is important to know the place value of each digit. The decimal point is pronounced as "and." For example, the number 333.3 is pronounced as "three hundred thirty-three and three tenths."

The diagram below will help you visualize the value of digits in each position.

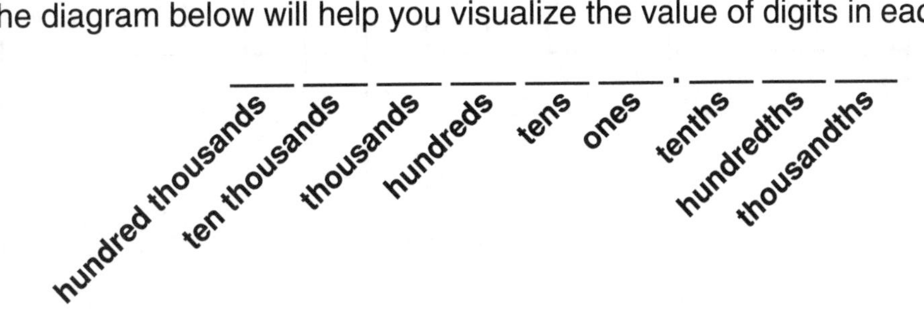

Quiz

Complete the following problems by placing the correct number by the correct place value.

Example: 78.1 = __**7**__ tens __**8**__ ones __**1**__ tenth

1. 14 = _____ten _____ones
2. 21.395 = _____tens _____ones _____tenths _____hundredths _____thousandths
3. 3.123 = _____ones _____tenths _____hundredths _____thousandths
4. 189.08 = _____hundreds _____tens _____ones _____tenths _____hundredths
5. 123 = _____hundreds _____tens _____ones
6. 0.39 = _____ones _____tenths _____ hundredths
7. 7,654 = _____thousands _____hundreds _____tens _____ones
8. 10,986 = _____ten thousands _____thousands _____hundreds _____tens _____ones
9. 9,045.7 = _____thousands _____hundreds _____tens _____ones _____tenths
10. 94,583 = _____ten thousands _____thousands _____hundreds _____tens _____ones

Name ______________________________ Date __________________

Place Value Crossword Puzzle

Complete the following crossword puzzle using the clues below.

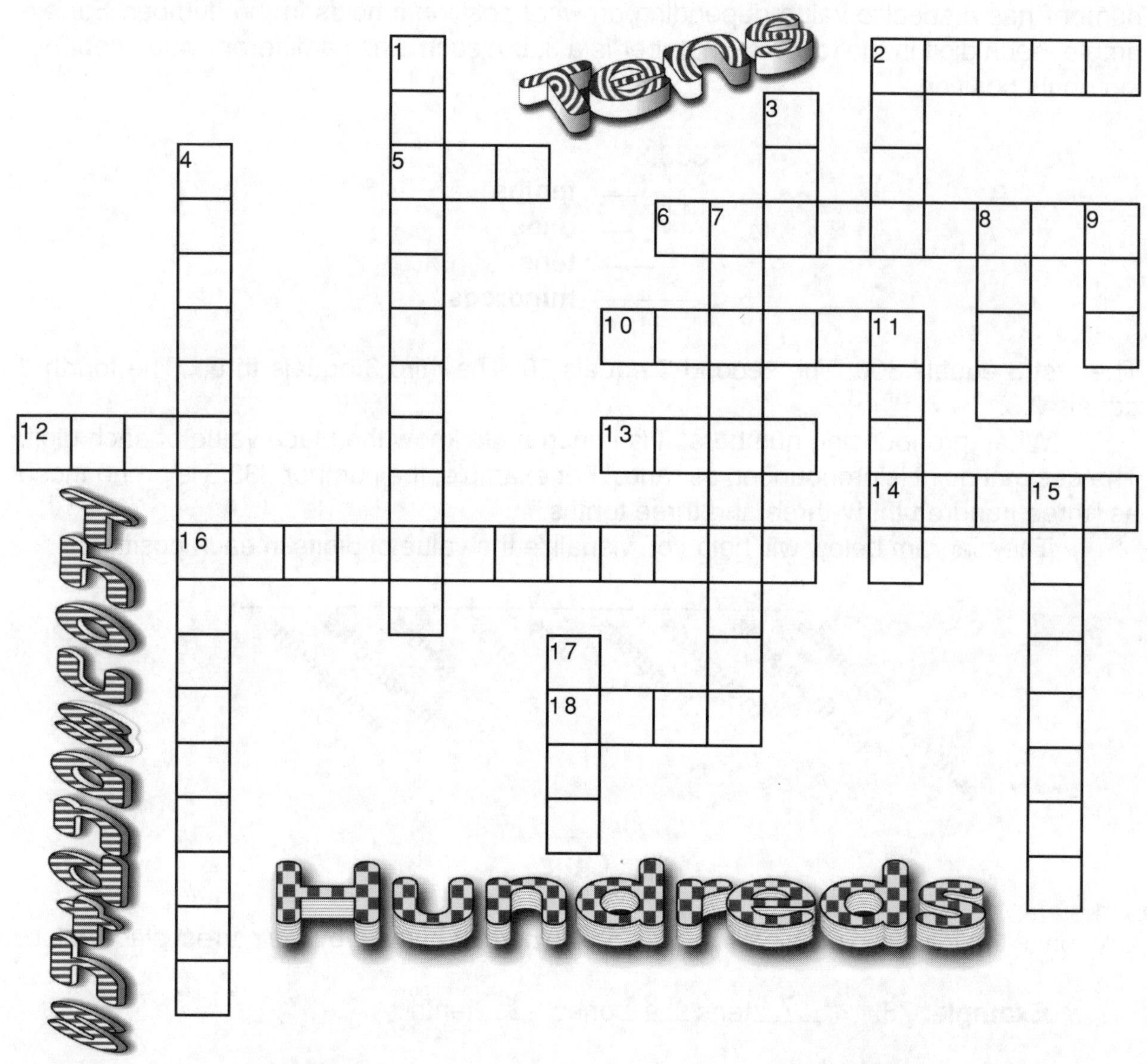

ACROSS

2. What number is in the hundreds place? 6,342.5
5. What number is in the hundredths place? 29.617
6. What place is the boldface number? 6**1**,847
10. What place is the boldface number? 55.**5**
12. What number is in the thousandths place? 2,179.045
13. What number is in the thousands place? 90,641
14. What number is in the hundred thousands place? 801,052.724
16. What place is the boldface number? 7**4**9,810.6
18. What place is the boldface number? 9**5**.21

DOWN

1. What place is the boldface number? 0.97**5**
2. What place is the boldface number? 2**6**8.01
3. What number is in the tens place? 1,329.06
4. What place is the boldface number? **1**16,522
7. What place is the boldface number? 0.8**6**4
8. What number is in the hundredths place? 0.298
9. What number is in the thousands place? 66,108
11. What number is in the ones place? 307
15. What place is the boldface number? 8,**1**29.96
17. What number is in the tenths place? 37.42

Name ______________________________ Date ________________

Addition

Addition involves the adding of two or more numbers (called **addends**) to get a resulting number called the **sum**. Addition problems are lined up vertically according to place value. (Decimal points must line up when doing addition problems.)

Examples:

$$\begin{array}{r} 6.2 \\ +\ 3 \\ \hline 9.2 \end{array} \qquad \begin{array}{r} 54 \\ +\ 32 \\ \hline 86 \end{array} \qquad \begin{array}{r} 83.5 \\ +\ 2.37 \\ \hline 85.87 \end{array}$$

The simplest type of addition is called **no renaming** because the sum of the numbers never exceeds nine, so there is no renaming or carrying involved.

Examples:

$$\begin{array}{r} 104 \\ +\ 281 \\ \hline 385 \end{array} \qquad \begin{array}{r} 24.7 \\ +\ 75.28 \\ \hline 99.98 \end{array} \qquad \begin{array}{r} 3284 \\ +\ 5114 \\ \hline 8398 \end{array}$$

Another type of addition is called **renaming**, because the sum of the numbers in a column equals ten or greater. This means the number in the tens place is then carried to the next column.

Examples:

$$\begin{array}{r} 1 \\ 174 \\ +\ 281 \\ \hline 455 \end{array} \qquad \begin{array}{r} 1\,1\,1 \\ 256.8 \\ +\ 383.4 \\ \hline 640.2 \end{array} \qquad \begin{array}{r} 1\ \ 1 \\ 3879 \\ +\ 7213 \\ \hline 11092 \end{array}$$

Quiz

Complete the following addition problems.

1. $\begin{array}{r} 22 \\ +\ 54 \\ \hline \end{array}$

2. $\begin{array}{r} 6.2 \\ +\ 2.7 \\ \hline \end{array}$

3. $\begin{array}{r} 27 \\ +\ 35 \\ \hline \end{array}$

4. $\begin{array}{r} 17.3 \\ +\ 28.6 \\ \hline \end{array}$

5. $\begin{array}{r} 534 \\ +\ 167 \\ \hline \end{array}$

6. $\begin{array}{r} 37.28 \\ +\ 21.22 \\ \hline \end{array}$

7. $\begin{array}{r} 2003 \\ +\ 482 \\ \hline \end{array}$

8. $\begin{array}{r} 523.88 \\ +\ 227.91 \\ \hline \end{array}$

9. $\begin{array}{r} 3872 \\ +\ 2383 \\ \hline \end{array}$

Name ______________________________ Date ________________

Addition Magic Squares

The squares below are "magic squares." They are magic because each horizontal and diagonal row and each vertical column will equal the same number when added. Find the numbers that fit in each row and column, as well as the "magic number" (the sum when each row or column is added). The first one is done for you. Using the blank square, create your own magic square and have your classmates find the magic number. (Hint: To create your own square, you must discover the pattern that makes the magic squares work.)

6	(11)	4
5	7	(9)
(10)	3	8

Magic Number 21

7	26	
	18	
15	10	29

Magic Number ________

64		141
	93	16
45		122

Magic Number ________

88		
	156	101
169		224

Magic Number ________

	107	90
	73	
56		124

Magic Number ________

	5	
33		49
	77	19

Magic Number ________

		134
	115	
96	170	79

Magic Number ________

68		272
		17
170	119	

Magic Number ________

Magic Number ________

Name ______________________ Date ____________

Addition Word Search

Complete the following addition problems. Then find and circle in the word search the missing word(s) for the clues below. The clues are the answers to the addition problems.

1. 82 + 25	2. 357 + 32	3. 59.7 + 10.9	4. 8945 + 345
5. 456 + 545	6. 658.6 + 37.6	7. 441.2 + 764.2	8. 2315 + 4197
9. 5678 + 12454	10. 72.86 + 72.86	11. 2165 + 764	12. 69.476 + 45.85

1. ______ ______ Seven
2. Three hundred ______-______
3. ______ ______ ______ tenths
4. Nine thousand ______ ______ ______
5. One ______ ______
6. ______ ______ ______-______ and two tenths
7. One thousand two hundred ______ ______ ______ ______
8. Six thousand ______ ______ ______
9. Eighteen thousand ______ ______ ______-______
10. One hundred forty-five and ______-______ ______
11. ______ ______ ______ hundred twenty-nine
12. One hundred fifteen and three hundred ______-______ ______

```
P J T T F P A J K R S Y E B H E I T S P R S X O
M F W R Z B S E V E N T Y A N D S I X W E E D N
U H E V D W O W E G T U G B M I G Y S W P V L E
G H N J X B K T S L P R I T N Q U J J R F E S H
X M T U U T W O T H O U S A N D N I N E V N Z U
I L Y X R Z T R H I G C K P N H V L L F R T F N
S T - W U A Z C L Y O O J V C G D L Q I Y Y B D
- R S T Y T E N I N D E R D N U H O W T Z - L R
Y L I J E J H A F X N Q R V Z A J P R C O T E E
T O X H N A F T N J L Q K D P Y E G J E T W Y D
E U T G V A M G G T T B E O X N D M I O L O W T
N H H L G A E D N C T R N V O C E G X X A H R H
I S O H X K K B X J D S H D M N H W I B R U S I
N L U C F V U P U N E S N T G T D W N D U N Z R
D T S P S Q B X U Q L A X G Y T C D O N P D U T
E K A A T G D H S K S K M - V L O Q E X W R D Y
R H N Q S N E A H U K P N J F T G Y H K O E R -
D F D W G N I L O M C I W D R T D U B M Y D M T
N F T X O Y M H S G N H R R J Q A C G Z F T R W
U I H W S Z T M Q E L K X Z L D N T U M O H J O
H T S K E L X P P B M R O R H X R Z K W K S J O
X M N V I F I V E A N D F O U R T E N T H S D L
I C W O C P D U O U J P L C S V O X B N D D K P
S X S J N E V L E W T D E R D N U H E V I F W W
```

Name ______________________________ Date ____________________

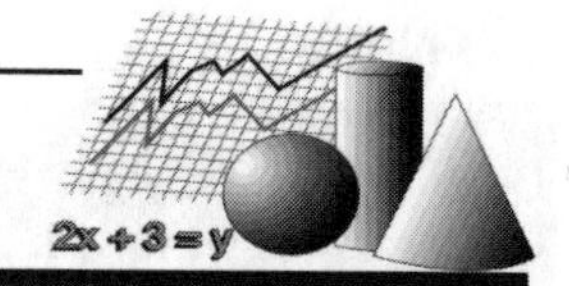

Addition Crossword Puzzle

Complete the following crossword puzzle using the clues given below. Include necessary decimal points in the answers.

ACROSS

1. What is the sum of 87.9, 6.73, and 4.08?
3. What is the total of 20,396 and 10,672?
5. Find the sum of 1,523.6 and 38.1.
7. Put together 727, 351, 490, and 118.
9. Combine 37,289 and 15,721.
12. Put together 82, 93, and 87.
14. Find the sum of 1,268; 722; and 2,542.
15. Combine 0.14, 26.7, 6.032, and 28.
17. Total the following numbers: 0.07, 0.28, 0.14, 0.007, and 0.014.
18. Find the sum of 8.9, 4.7, and 63.8.
20. What is the total of 323 and 213?
21. Combine 209 and 109.
23. Add 327, 123, 970, and 620.

DOWN

2. Combine 8,746 and 96,487.
4. How much is 4.27, 16.28, 47.92, and 32.59 added together?
5. Add 268.13, 19.677, and 1345.43.
6. What is the total of 67.1 and 5.9?
7. What is the sum of 268, 783, and 421?
8. What is the total of 168, 947, and 89?
10. Find the sum of 477 and 376.
11. Add 26, 82, and 19.
13. Put together 8,762; 6,241; and 6,565.
16. Combine 3,563; 1,789; and 2,511.
19. Find the sum of 296.83, 65.897, 246.11, and 156.42.
22. Add 5,321 and 2,998.

Name ______________________________ Date ________________

Subtraction

Subtraction is the process of finding the **difference** of two numbers. The **minuend** is the number being subtracted from, and the **subtrahend** is the number subtracted from the minuend.

The simplest type of subtraction is called **no renaming** (**no borrowing**) because the subtrahend is smaller than the minuend and can be subtracted from the minuend.

Examples:

$$\begin{array}{r} 98 \\ -\ 35 \\ \hline 63 \end{array} \qquad \begin{array}{r} 853.7 \\ -\ 241.5 \\ \hline 612.2 \end{array} \qquad \begin{array}{r} 4756 \\ -\ 2133 \\ \hline 2623 \end{array}$$

Another type of subtraction is called **renaming** or **borrowing** because the subtrahend is larger than the minuend, so ten must be borrowed from the number to the left and added to the minuend before the subtraction can take place.

Examples:

$$\begin{array}{r} 5\ 12 \\ \not6\not2 \\ -\ 9 \\ \hline 53 \end{array} \qquad \begin{array}{r} 3\ 12 \\ 2\not4\not2 \\ -\ 163 \\ \hline 9 \end{array} \qquad \begin{array}{r} 1\ 13 \\ \not2\not4\not2 \\ -\ 163 \\ \hline 79 \end{array}$$

Quiz

Complete the following subtraction problems.

1. $\begin{array}{r} 89 \\ -\ 68 \\ \hline \end{array}$

2. $\begin{array}{r} 6.2 \\ -\ 2.7 \\ \hline \end{array}$

3. $\begin{array}{r} 372 \\ -\ 251 \\ \hline \end{array}$

4. $\begin{array}{r} 125 \\ -\ 98 \\ \hline \end{array}$

5. $\begin{array}{r} 867 \\ -\ 654 \\ \hline \end{array}$

6. $\begin{array}{r} 37.28 \\ -\ 21.22 \\ \hline \end{array}$

7. $\begin{array}{r} 4568 \\ -\ 3569 \\ \hline \end{array}$

8. $\begin{array}{r} 523.88 \\ -\ 227.91 \\ \hline \end{array}$

9. $\begin{array}{r} 3872 \\ -\ 2383 \\ \hline \end{array}$

10. $\begin{array}{r} 3429.24 \\ -\ 2367.43 \\ \hline \end{array}$

11. $\begin{array}{r} 52437 \\ -\ 31859 \\ \hline \end{array}$

12. $\begin{array}{r} 245.38 \\ -\ 12.988 \\ \hline \end{array}$

Name ______________________________ Date ________________

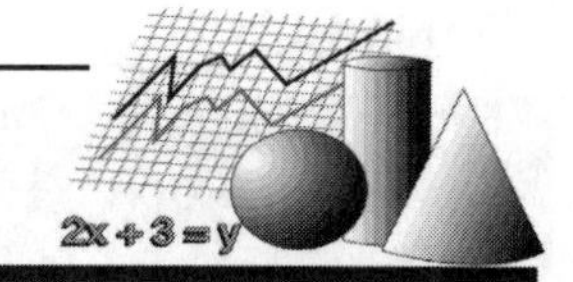

Subtraction Crossword Puzzle

Complete the following crossword puzzle using the clues given below. Include necessary decimal points in the answers.

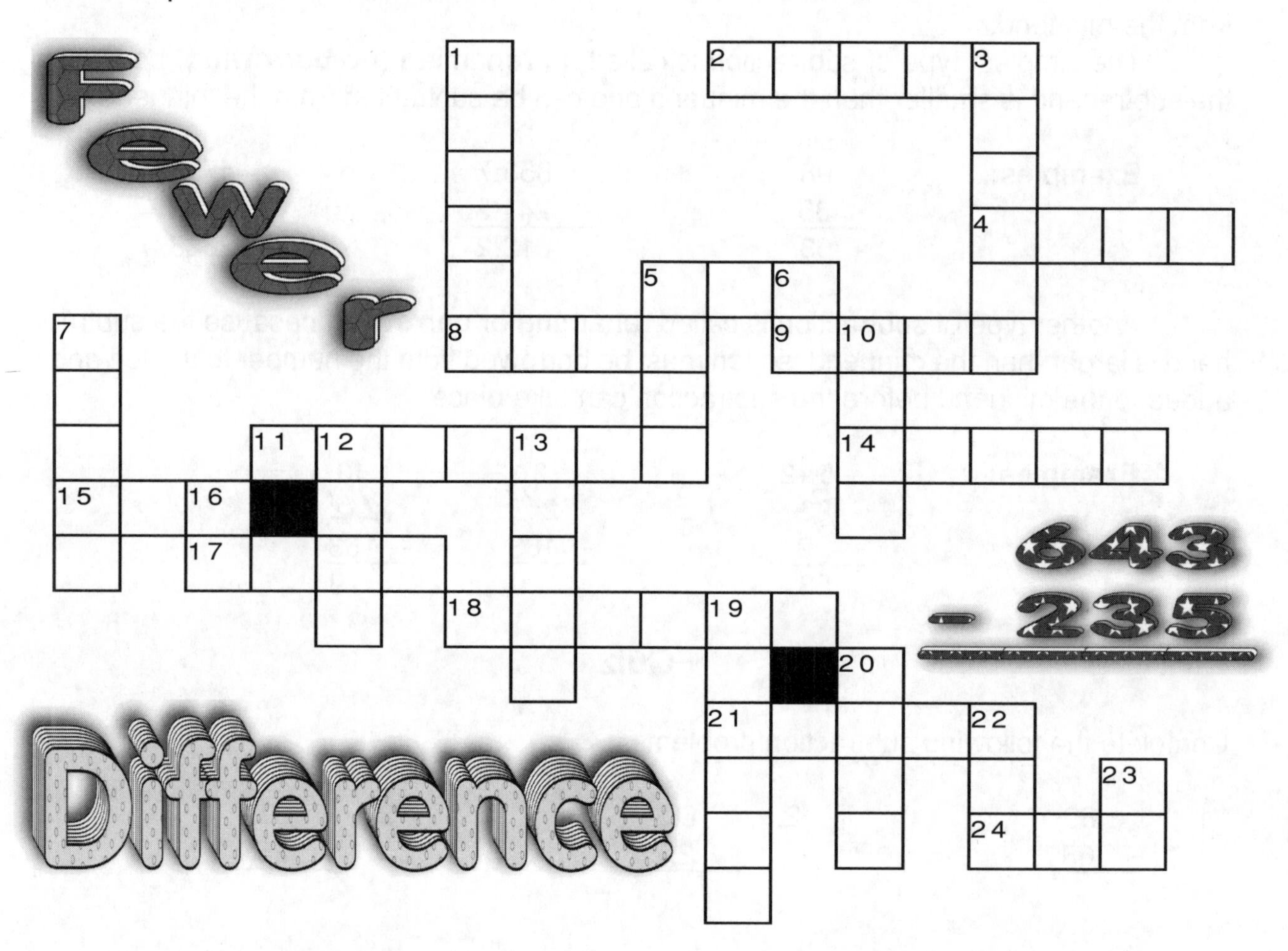

ACROSS

2. How much less is 8,829 than 71,921?
4. What's the difference between 7,549 and 1,909?
5. How many fewer is 486 than 673?
8. How much less is 6.4 than 10.01?
9. How much less is 309 than 4,593?
11. How much more is 68.05 than 4.0372?
14. How much less is 0.001 than 0.01?
15. How much more do you need to go from 637 to 800?
17. How much more is 15,384 than 6,154?
18. What amount is needed to go from 13.25 to 1,000?
21. Subtract 17.58 from 81.9.
24. How many fewer is 2,715 than 3,679?

DOWN

1. Subtract 182,838 from 432,651.

DOWN (continued)

3. Subtract 1.2595 from 3.6149.
5. Subtract 8.2 from 20.
6. Subtract 89 from 163.
7. How much more is 17,008 than 5,991?
10. What amount is needed to go from 4,327 to 7,234?
12. What is the difference between 9,118 and 4,381?
13. What amount is needed to go from 82.16 to 100?
16. How much more do you need to go from 39 to 78?
19. What is the difference between 826.4 and 59.63?
20. How much more do you need to go from 6.29 to 10?
22. What's the difference between 517 and 268?
23. How much less is 28 than 82?

Name ______________________________ Date ________________

Subtraction Hidden Number Puzzle

Complete the subtraction word problems below and write the answers in the blanks (first the number form and then the word form). Then unscramble the circled blanks to answer the hidden number word problem.

1. Cindy has to read 2,500 pages this semester to be eligible for a free pizza. She read 507 pages the first week, 674 the second week, and 469 the third week. How many pages must she read the fourth week to win the free pizza? ________

   ___ ___ ___ ___ ___ ___ (___) ___ ___ ___ ___ ___ ___ (___)

2. The demolition of a 124-story skyscraper must be done in stages because of how close it is to other buildings. The demolition took 22 stories in the first stage and 26 stories in the second stage. How many stories remain to be demolished? ________

   (___) ___ ___ ___ ___ ___ ___ - ___ ___ ___

3. Ben was selling lemonade at his street stand. He began the day with 256 ounces of lemonade. He sold three 16-ounce glasses the first hour, four 12-ounce glasses the second hour, and six 8-ounce glasses the third hour. How many ounces does he have left? ________

   ___ ___ ___ ___ ___ ___ ___ ___ ___ ___ ___ ___ (___) ___ ___ ___

4. Verne is putting a fence around a square lot that is 37.75 feet on each side. He has 225 feet of fencing material. How much material will Verne have left after going around the square lot?

   ________ ___ ___ ___ ___ ___ ___ ___ ___ - ___ (___) ___ ___

5. Isabelle rides the subway each day. She gets on at the 128th Street entrance. Then she rides to 39th Street. How many blocks has she gone? ________

   ___ ___ ___ ___ ___ ___ (___) - ___ ___ (___) ___

6. Tony is an attendant at a parking lot. He has 350 spaces in the lot. He allowed 75 vehicles to park in each of the first three hours. How many spaces are left? ________

   ___ ___ (___) ___ ___ ___ ___ ___ ___ ___ ___ ___ ___ ___ ___ ___ ___ - ___ ___ (___) ___

7. Karen was canning vegetables. She started with 75 quart jars. She used 15 the first night, 25 the second night, and 19 the third night. How many jars does she have left?

   ________ ___ ___ ___ ___ (___) ___ ___

Hidden Number: Try to unscramble the hidden number first. If you can't figure it out, determine the sum of the answers in questions four and five. Then determine the sum of the answers in questions two and seven. Then subtract the smaller number from the larger number.

Hidden Number: ___ ___ ___ ___ ___ ___ ___ ___ - ___ ___ ___

Name ______________________________ Date ________________

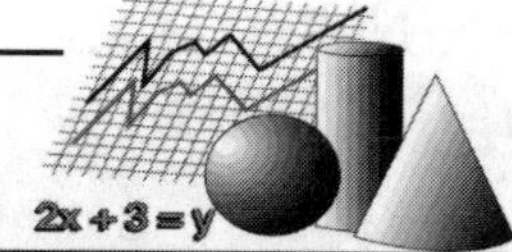

Subtraction Hidden Picture Puzzle

80.2	179	378	5.672	45	983	2074	211	90	6501
425	881	7.64	1.06	12.98	616	7053	0.418	2573	194
35	827	952	512	7.32	24.09	532	906	3356	87.42
135	17.62	7.54	139	90.3	267	5223	431	1.95	82.08
569	293	3602	4576	8.6	924	7645	1.86	3.554	6.22
12.34	61.52	487	6.35	9467	3002	88.4	728	1.342	39.5
46	82.67	287.3	14.12	790	18.63	6054	1.683	952	10.87
34.90	726	147	109	5.832	67.20	2.39	2603	10.15	521
799	1462	555	14.09	9.311	8.6	3072	195.2	898	633.2
385	4.754	78.2	643	153.9	9051	76.1	383.6	62	14.13

Solve the problems below. Then find each answer in the grid above and shade the square in which the answer appears to reveal a hidden picture.

1. 27,482 - 22,906 = ____________
2. 1,601 - 875 = ____________
3. 9,733 - 4,510 = ____________
4. 561 - 74 = ____________
5. 14,532 - 11,929 = ____________
6. 82,633 - 81,905 = ____________
7. 749 - 318 = ____________
8. 2,598 - 2,459 = ____________
9. 68,425 - 68,132 = ____________
10. 657 - 548 = ____________
11. 8.52 - 6.57 = ____________
12. 74.20 - 60.08 = ____________
13. 97.51 - 79.89 = ____________
14. 351.20 - 341.05 = ____________
15. 4.83 - 2.97 = ____________
16. 90.5 - 7.83 = ____________
17. 4.86 - 2.47 = ____________
18. 10,160.42 - 10,098.9 = ____________
19. 5.403 - 3.72 = ____________
20. 30.50 - 24.15 = ____________

Name ______________________________ Date ______________

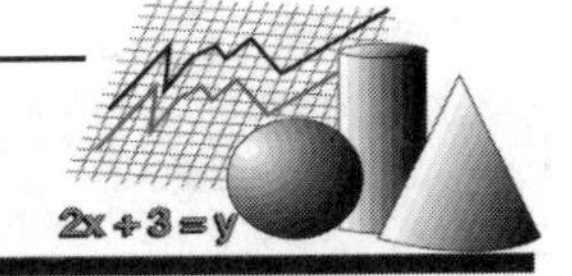

Multiplication

Multiplication is the process of finding the **product** of two numbers called **factors**.

In multiplication of factors of two digits or larger, each digit in the second factor is multiplied by each digit in the first factor, and then the products are added together to get the final product.

If the product of the multiplication of two digits is greater than ten, the number in the tens place is carried to the next digit to the left, and it is added to the product of the next multiplication.

Examples:

$$\begin{array}{r} {}^{2}\\ 78 \\ \times\ 3 \\ \hline 234 \end{array} \qquad \begin{array}{r} {}^{3}\\ 24 \\ \times\ 8 \\ \hline 192 \end{array} \qquad \begin{array}{r} {}^{3}\\ 67 \\ \times\ 5 \\ \hline 335 \end{array}$$

$$\begin{array}{r} 413 \\ \times\ 271 \\ \hline 413 \end{array} \qquad \begin{array}{r} {}^{2}\\ 413 \\ \times\ 271 \\ \hline 413 \\ 2891 \end{array} \qquad \begin{array}{r} 413 \\ \times\ 271 \\ \hline 413 \\ 2891 \\ 826 \end{array} \qquad \begin{array}{r} 413 \\ \times\ 271 \\ \hline 413 \\ 2891 \\ 826 \\ \hline 111923 \end{array}$$

Quiz

Complete the following multiplication problems.

1. $\begin{array}{r} 27 \\ \times\ 4 \\ \hline \end{array}$

2. $\begin{array}{r} 5.4 \\ \times\ 0.8 \\ \hline \end{array}$

3. $\begin{array}{r} 59 \\ \times\ 34 \\ \hline \end{array}$

4. $\begin{array}{r} 42.2 \\ \times\ 22.4 \\ \hline \end{array}$

5. $\begin{array}{r} 763 \\ \times\ 45 \\ \hline \end{array}$

6. $\begin{array}{r} 1314 \\ \times\ 15 \\ \hline \end{array}$

7. $\begin{array}{r} 2510 \\ \times\ 123 \\ \hline \end{array}$

8. $\begin{array}{r} 551 \\ \times\ 456 \\ \hline \end{array}$

9. $\begin{array}{r} 38.72 \\ \times\ 6.45 \\ \hline \end{array}$

Name ______________________________ Date ____________________

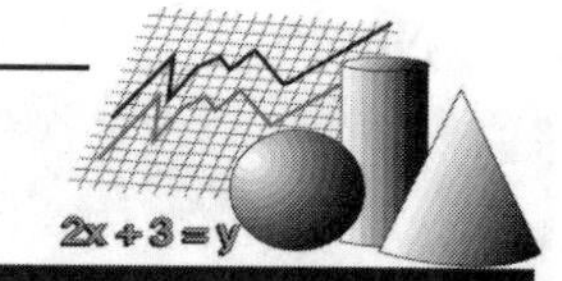

Multiplication Chart

Complete the following chart.

X	1	2	3	4	5	6	7	8	9	10	11	12	13	14	15
1															
2															
3															
4															
5															
6															
7															
8															
9															
10															
11															
12															
13															
14															
15															
16															
17															
18															
19															
20															

To complete the chart:

Come down the left column under "x" until you reach 3. Move across the row 3 is on until you are directly under 2 in the top row. Multiplying 3 x 2 = 6. So a 6 goes in the square across from 3 and under 2. (You can make a chart using negative numbers as well. Remember that multiplying two negatives equal a positive and multiplying one negative and one positive equal a negative.) Use this procedure and complete the remaining squares.

The numbers in the shaded squares are perfect squares.

Name ______________________________ Date ________________

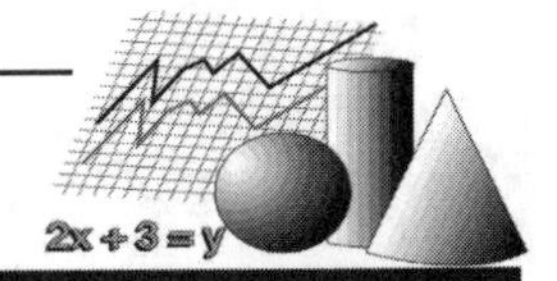

Multiplication Crossword Puzzle

Complete the following crossword puzzle using the clues given below. Include necessary decimal points in the answers.

ACROSS

2. Multiply 341 and 218.
7. 56 groups of 54.21
8. The product of 5,034 and 908
9. The product of 192 and 8
11. 72 rows of 3
13. 17 groups of 28
15. 326 groups of 1.17
18. 14 groups of 12
20. The product of 5.17 and 2.6
22. 52 rows of 15
23. Multiply 3.145 and 36.2.

DOWN

1. Multiply 776 and 8.
3. The product of 68 and 52
4. The product of 1.92 and 48
5. 911 groups of 6
6. 1,751 rows of 7
10. 46 rows of 12
12. The product of 39 and 16
13. 6 groups of 78
14. The product of 21 and 3
16. The product of 4.4 and 3.8
17. 8 groups of 25
18. 26 rows of 745
19. Multiply 55 and 15.
21. 938 groups of 45

Name ______________________________ Date ________________

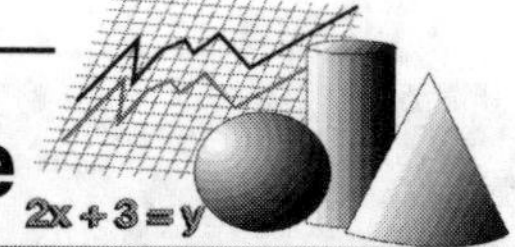

Multiplication Hidden Number Puzzle

Complete the multiplication word problems below and write the answers in the blanks (first the number form and then the word form). Then unscramble the circled blanks to answer the hidden number word problem.

1. Stacey is inventorying her store's candy bars. She notices that she has 17 cases with 24 candy bars in each. How many candy bars does she have? ________

   ___ ___ ___ ___ ___ ___ ___ ___ ___ ___ ___ ___ ___ ___ ___ ___ ___

2. The school's pencil dispenser has just been loaded with a gross of new pencils. The school sells 22 pencils each day, Monday through Friday. How many pencils were sold this week? ________

   ___ ___ ___ ___ ___ ___ ___ ___ ___ ___ ___ ___ ___ ___

3. Chip just purchased 25 disks. Each disk can handle 1.44 megabytes of data. How many megabytes of data will all of his disks hold? ________

   ___ ___ ___ ___ ___ - ___ ___ ___

4. Hillary has a garden with 12 tomato plants, and each plant produces 57 tomatoes. How many tomatoes does Hillary have? ________

   ___ ___ ___ ___ ___ ___ ___ ___ ___ ___ ___ ___ ___ ___ ___ ___ - ___ ___ ___ ___

5. Jessie rides the subway to and from school each day. It is 29 blocks to the school from her building. How many blocks does she ride the subway to and from school in a five-day week? ________

   ___ ___ ___ ___ ___ ___ ___ ___ ___ ___ ___ ___ ___ ___ ___ ___

6. Saul reads an average of 15 books a quarter. Each book contains about 204 pages. How many pages does Saul read in a quarter? ________

   ___

7. Harry surveys the local gas station. He finds out that the station averages 51 vehicles per day and each vehicle gets about 12 gallons of gas. How many gallons of gas are sold on a typical day? ________

   ___ ___ ___ ___ ___ ___ ___ ___ ___ ___ ___ ___ ___ ___ ___ ___

Hidden Number: Try to unscramble the hidden number first. If you can't figure it out, determine the product of the answers in questions two and five. Then determine the product of the answers in questions one and three. Subtract the smaller number from the larger number, and then subtract 16,800 from that number.

Hidden Number: ___

Name ______________________________ Date ________________

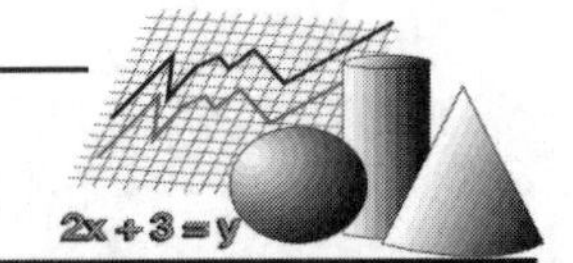

Division

Division is the process of finding how many times a number (the **divisor**) is contained in another number (the **dividend**). The answer in a division problem is called the **quotient**. For example, in the problem 13 ÷ 3, the 13 is the dividend and the 3 is the divisor. The goal is to find how many 3's are in 13.

Examples:

```
   4R1  ← quotient          7R2
3 )13   ← dividend       5 )37
   12                        35
↗  --                        --
divisor 1                     2
```

There are four 3's in 13 with a **remainder** of one in the first problem. The remainder is what is left over after the divisor is evenly divided into the dividend. It becomes part of the quotient.

To divide so that there are no remainders, decimal places must be added to the right of the dividend.

Example:

```
     4           4.5
4 )18       4 )18.0
   16          16
   --          --
    2           20
                20
                --
                 0
```

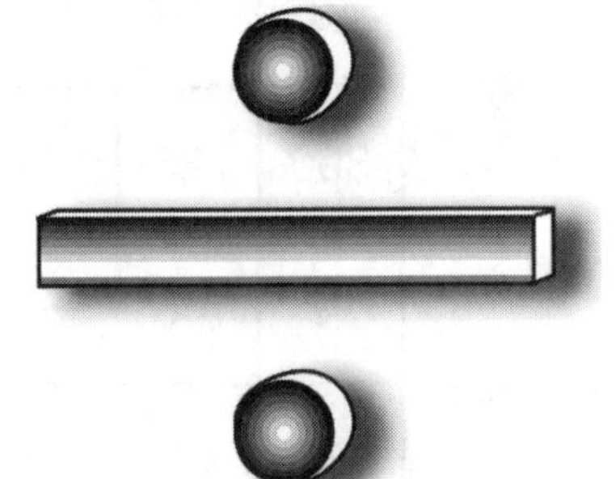

Quiz

Complete the following division problems. Write the answers for problems 1–2 in remainder form; write the answers for problems 3–6 in decimal form, and round to the nearest hundredth if required.

1. 5)19
2. 15)232
3. 76)665
4. 24)1584
5. 62)1521
6. 68)1105

Name ______________________ Date ______________

Division Hidden Picture Puzzle

80	82	378	672	45	983	73	211	90	651
425	881	21	106	298	707	53	41	573	194
35	87	952	417	212	29	52	906	31	84
51	107	361	867	56	19	17	689	195	88
69	293	362	43	44	24	745	186	354	22
123	65	98	66	967	63	84	728	142	39
46	839	287	112	70	183	34	168	952	87
390	726	147	109	58	67	23	263	101	521
99	146	555	14	93	86	372	152	89	633
385	47	72	643	153	951	71	38	62	141

Solve the problems below. Then find each answer in the grid above and shade the square in which the answer appears to reveal a hidden picture.

1. 4,234 ÷ 58 = ______________
2. 1,292 ÷ 68 = ______________
3. 2,415 ÷ 115 = ______________
4. 22,274 ÷ 518 = ______________
5. 4,815 ÷ 45 = ______________
6. 17,384 ÷ 82 = ______________
7. 177,029 ÷ 211 = ______________
8. 6,363 ÷ 9 = ______________
9. 4,200 ÷ 75 = ______________
10. 544 ÷ 16 = ______________
11. 54,431 ÷ 79 = ______________
12. 2,952 ÷ 36 = ______________
13. 9,537 ÷ 11 = ______________
14. 1,188 ÷ 27 = ______________
15. 1,683 ÷ 33 = ______________
16. 80,064 ÷ 192 = ______________
17. 3,655 ÷ 215 = ______________
18. 882 ÷ 9 = ______________
19. 24,187 ÷ 67 = ______________
20. 53,487 ÷ 849 = ______________

Name ______________________________ Date ________________

Division Crossword Puzzle

Complete the crossword puzzle by completing the word problems given below. Include necessary decimal points in the answers.

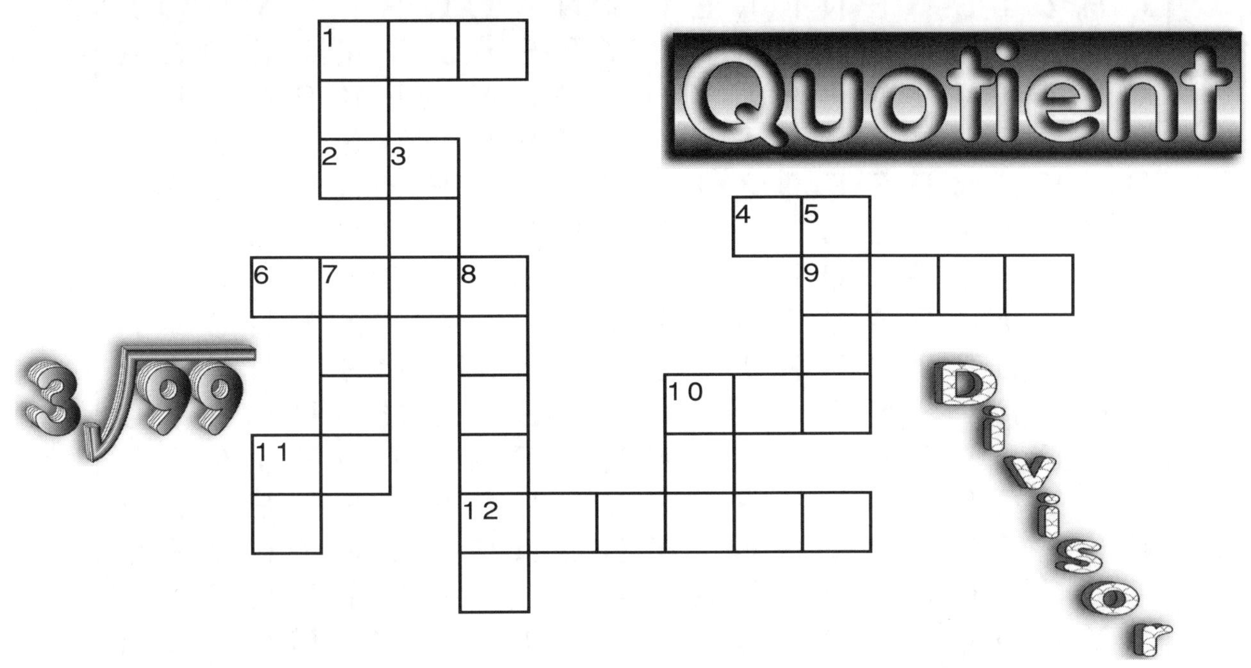

ACROSS

1. Tom rushed to 1,260 yards in the nine games he played this season in football. How many yards did he average in each game?
2. Brian's father divided his baseball card collection of 324 cards among his four children. How many cards did each child get?
4. James estimates that he walks at least 258 blocks each week. If he walks six days a week, how many blocks will he walk each day?
6. There are 31,992 fans at a baseball game. There are 24 main sections of seating in the baseball park. How many people are sitting in each section if each section is filled up equally?
9. You have 10,000 candy bars to sell for a fund raiser, and four grades among which to equally split them. How many candy bars will each grade have to sell?
10. A staircase is 68 inches high, and each step is eight inches high. How many steps can you get on the staircase?
11. A school has 175 students in it. There are seven grades in the school and each grade has the same number of students in it. How many students are in each grade?
12. A computer's hard drive is 640,000,000 bytes in size. If each kilobyte is 1,024 bytes, how many kilobytes are in the computer's hard drive?

DOWN

1. The school dance committee has 752 cans of soda for the students at a dance. How many students are at the dance if there are enough cans of soda for everyone to have four?
3. Holly must read 2,860 pages this month. She plans to read the same number of pages for 20 nights. How many pages must she read each night?
5. A recordable CD can hold 650 megabytes of data. You can put 20 programs on each CD. How many megabytes must each program be for each to fit equally on the CD?
7. If a punch bowl can hold 244 ounces of punch, how many eight-ounce glasses of punch can be dipped from the punch bowl?
8. Cassidy had a garden that produced 550 potatoes. She had 15 plants. How many potatoes were produced by each plant? (Round to the nearest thousandth.)
10. A gas station has a 20,000-gallon tank of gas. How many vehicles can the tank service if each vehicle uses 25 gallons of gas?
11. Jerome helps his brother clean offices after school. The building he is in has 360 offices and there are 18 floors. How many offices are on each floor on average?

Name ______________________________ Date ____________________

Division Word Search

```
O E V L E W T E H U F B Q I O V E I G H T E E N
Y W D Q O B B K Q Z S J G W X B V C A C B C P C
I H Z J S I E N I N E F T N A Q Q R I V Y P Q X
M S E Y E C E P G A F - Y P S I X T Y - F O U R
V I T X V V V R H Z Y H V O P Y W P K P G C H L
P X M Q E W F E H T B N O V V N I D S R D F E P
D T W K N F E F R Z J C H O I D F X A X O B P E
G Y R T T Z B O X K F E T N B Z I H C R R C N O
C - U Q Y G F U I S D N E R Y S N B T E A I T S
L T T W - R A Y J N U T J I - L C Y X M N C V I
K W R U F W Y U R J Y L J Y Y J - M S - H X C Y
T O T A I H V Q Q - T J T T B F B N Y U L R T Q
H E E S V O N Y F L V N N Y O G N T U Z J N R X
I B V O E O H I Z R E C G U L B H Z E I E G P H
R L I E Y Z V V G W T Y R C N G A J B W N V R O
T Z F I Q E V M T U A K U P I Q T I T D R O P W
Y Y K D E Q X D N U P G A E J Y E W L M I T M H
- X T H R E E A N S C U X F E N I N - Y T R O F
T M K D E A J E K J D Q R Q R X C Y L H H N Q J
W E B O N E H U N D R E D N I N E T Y Y Z F K A
O N K K B P X C C S U E X Z G R H P D A J C U R
S D G F I V E H U N D R E D F I F T Y - S I X E
M V O I K J Z P A E A P D Y M U A L E K M Q A I
R U O F L Q M F Z N E V E S - Y T X I S J J X V
```

Solve the problems below. Then find and circle each answer in the word search above.

1. 182 ÷ 7 = ____________
2. 672 ÷ 16 = ____________
3. 5,270 ÷ 85 = ____________
4. 356 ÷ 4 = ____________
5. 486 ÷ 54 = ____________
6. 1,628 ÷ 37 = ____________
7. 2,660 ÷ 28 = ____________
8. 1,005 ÷ 15 = ____________
9. 1,792 ÷ 28 = ____________
10. 343 ÷ 7 = ____________
11. 384 ÷ 12 = ____________
12. 2,850 ÷ 15 = ____________
13. 230 ÷ 46 = ____________
14. 378 ÷ 21 = ____________
15. 144 ÷ 12 = ____________
16. 267 ÷ 89 = ____________
17. 140 ÷ 35 = ____________
18. 33,916 ÷ 61 = ____________
19. 1,080 ÷ 54 = ____________
20. 4,425 ÷ 59 = ____________

Name ______________________________ Date ______________

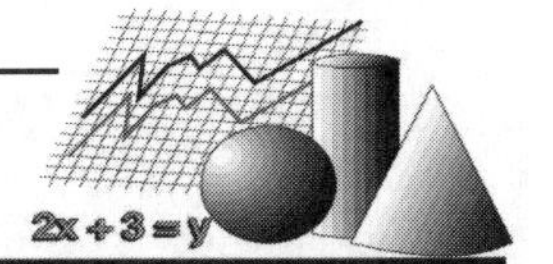

Fractions

A **fraction** expresses the part of the whole that is present. A fraction is made up of two parts, the numerator (the top number) and the denominator (the bottom number). The **numerator** tells how many parts are present. The **denominator** tells how many parts the whole is broken into.

denominator → $\frac{1}{2}$ ← **numerator**

A whole circle divided into four equal parts can be represented by the following fraction.

$\frac{4}{4}$ If one of the four parts is shaded, this can be represented by the following fraction. $\frac{1}{4}$ 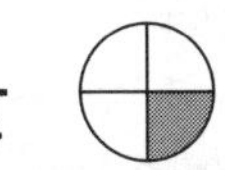

Fractions can be added and subtracted. To add or subtract a fraction, the denominator must be the same number. If the denominators are the same, then all you do is add or subtract the numerators. The answer is the addition or subtraction of the numerators written over the denominator.

Examples: $\frac{1}{5} + \frac{2}{5} = \frac{1+2}{5} = \frac{3}{5}$ $\frac{3}{7} - \frac{1}{7} = \frac{3-1}{7} = \frac{2}{7}$

If the denominators are not the same, we must make them the same. To do this we must find the **least common denominator** (LCD) for both fractions. The LCD is the smallest number that both denominators can divide into without decimals or remainders. This requires two steps.

Step 1: Determine if either denominator can divide into the other denominator. If so, the larger denominator is the LCD. If not, go to Step 2. You must convert both fractions so that the denominators are the same. This is called creating **equivalent fractions**. You must multiply the numerator by the same number you multiplied the denominator by.

Example: $\frac{2}{5} + \frac{3}{10} = \frac{2 \times 2}{5 \times 2} + \frac{3}{10} = \frac{4}{10} + \frac{3}{10} = \frac{4+3}{10} = \frac{7}{10}$

Step 2: The LCD is the product of both denominators.

Example: $\frac{4}{7} + \frac{5}{9} = \frac{4 \times 9}{7 \times 9} + \frac{5 \times 7}{9 \times 7} = \frac{36}{63} + \frac{35}{63} = \frac{71}{63} = 1\frac{8}{63}$

Multiplication of fractions is very simple. It does not matter what the denominators are; you just multiply the numerators together and multiply the denominators together.

Examples: $\frac{5}{8} \times \frac{3}{4} = \frac{5 \times 3}{8 \times 4} = \frac{15}{32}$ $\frac{4}{12} \times \frac{7}{8} = \frac{4 \times 7}{12 \times 8} = \frac{28}{96}$

Name ______________________________ Date ________________

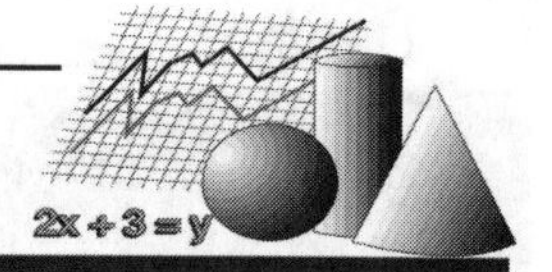

Fractions (continued)

Division of fractions is a little different. When you divide a fraction by a fraction, you must take the second fraction and invert it (meaning turn it upside down). Then change the division symbol to a multiplication symbol and multiply the two fractions. The answer is the final product.

Example: $\frac{1}{4} \div \frac{3}{5} = \frac{1}{4} \times \frac{5}{3} = \frac{1 \times 5}{4 \times 3} = \frac{5}{12}$

When the numerator is larger than the denominator, the fraction is called an **improper fraction**. Improper fractions must be rewritten as a mixed number. A **mixed number** is a whole number such as 1, 2, 3, and so on, with a proper fraction written next to it.

Examples: $\frac{12}{7} = 1\frac{5}{7}$ $\frac{15}{4} = 3\frac{3}{4}$ $\frac{23}{5} = 4\frac{3}{5}$

To convert a mixed number to an improper fraction, you must take the denominator and multiply it by the whole number portion of the mixed number. After you multiply these two numbers, you then add the numerator to the product and this becomes the new numerator for the improper fraction. The denominator will always remain the same.

Example: $2\frac{5}{11} = \frac{(2 \times 11) + 5}{11} = \frac{22 + 5}{11} = \frac{27}{11}$

The final consideration for fractions is whether the fraction is written in **simplest form**. The simplest form is when there is no number other than the number 1 that can be divided into both the numerator and denominator.

Examples: $\frac{2}{4} = \frac{2 \div 2}{4 \div 2} = \frac{1}{2}$ $\frac{12}{16} = \frac{12 \div 2}{16 \div 2} = \frac{6 \div 2}{8 \div 2} = \frac{3}{4}$

Quiz

Complete the following problems involving fractions.

1. $\frac{2}{8} + \frac{3}{8} =$
2. $\frac{5}{8} + \frac{1}{2} =$
3. $\frac{7}{14} - \frac{3}{14} =$
4. $\frac{9}{10} - \frac{1}{5} =$
5. $9\frac{1}{2} - 3\frac{2}{4} =$
6. $\frac{5}{8} \times \frac{2}{7} =$
7. $2\frac{5}{6} \times 3\frac{2}{9} =$
8. $\frac{4}{9} \div \frac{15}{16} =$
9. $5\frac{2}{7} \div \frac{1}{3} =$

Name ______________________ Date ______________

Fractions Crossword Puzzle

Complete the crossword puzzle using the clues given below. Hyphens are not included in the answers.

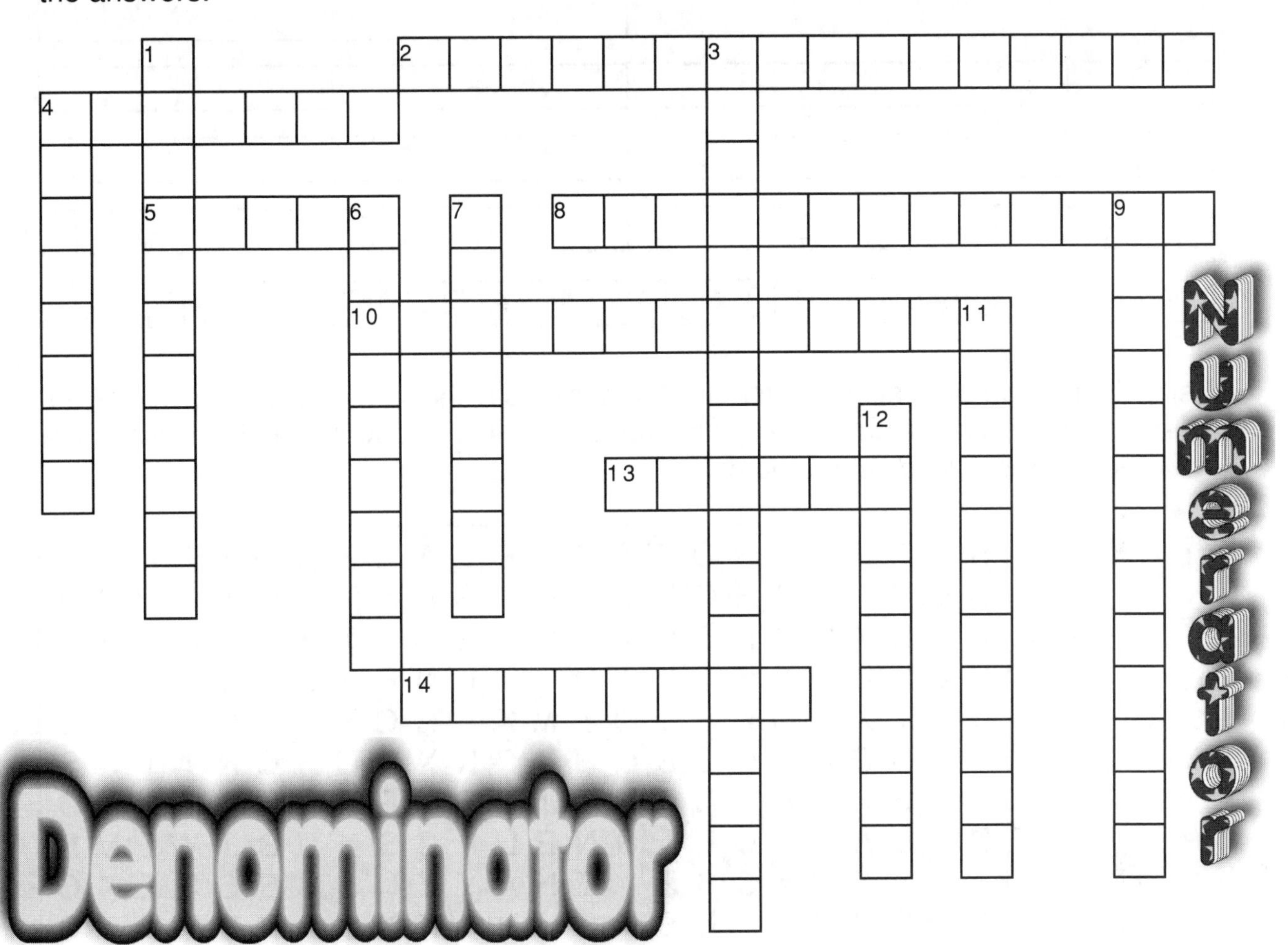

ACROSS

2. This fraction with a denominator of five equals the whole number five. (three words)
4. The fraction $\frac{7}{14}$ equals this simple fraction. (two words)
5. A common denominator for the fractions $\frac{1}{2}$ and $\frac{3}{4}$ (one word)
8. This fraction equals the mixed number $7\frac{1}{2}$ (two words)
10. This mixed number is three times as large as $\frac{1}{2}$. (four words)
13. A common denominator for the fractions $\frac{3}{4}$ and $\frac{5}{6}$ (one word)
14. When the numerator of the fraction is larger than the denominator, the fraction is called an _____ fraction.

DOWN

1. This simple fraction is equivalent to the fraction $\frac{14}{18}$. (two words)
3. This improper fraction with a denominator of seven equals the whole number seven. (three words)
4. The fraction $\frac{2}{12}$ equals this simple fraction. (two words)
6. This simple fraction is twice as large as $\frac{1}{3}$. (two words)
7. This simple fraction is twice as large as $\frac{1}{6}$. (two words)
9. This fraction equals the mixed number $2\frac{3}{4}$. (two words)
11. This simple fraction is equivalent to the fraction $\frac{10}{16}$. (two words)
12. This fraction with a numerator of 10 equals the whole number one. (two words)

Name ______________________________ Date ____________________

Fractions Word Search

Fraction Chart

$\frac{1}{2}$	$\frac{2}{2}$

$\frac{1}{4}$	$\frac{2}{4}$	$\frac{3}{4}$	$\frac{4}{4}$

$\frac{1}{8}$	$\frac{2}{8}$	$\frac{3}{8}$	$\frac{4}{8}$	$\frac{5}{8}$	$\frac{6}{8}$	$\frac{7}{8}$	$\frac{8}{8}$

$\frac{1}{16}$	$\frac{2}{16}$	$\frac{3}{16}$	$\frac{4}{16}$	$\frac{5}{16}$	$\frac{6}{16}$	$\frac{7}{16}$	$\frac{8}{16}$	$\frac{9}{16}$	$\frac{10}{16}$	$\frac{11}{16}$	$\frac{12}{16}$	$\frac{13}{16}$	$\frac{14}{16}$	$\frac{15}{16}$	$\frac{16}{16}$

$\frac{1}{32}$	$\frac{2}{32}$	$\frac{3}{32}$	$\frac{4}{32}$	$\frac{5}{32}$	$\frac{6}{32}$	$\frac{7}{32}$	$\frac{8}{32}$	$\frac{9}{32}$	$\frac{10}{32}$	$\frac{11}{32}$	$\frac{12}{32}$	$\frac{13}{32}$	$\frac{14}{32}$	$\frac{15}{32}$	$\frac{16}{32}$	$\frac{17}{32}$	$\frac{18}{32}$	$\frac{19}{32}$	$\frac{20}{32}$	$\frac{21}{32}$	$\frac{22}{32}$	$\frac{23}{32}$	$\frac{24}{32}$	$\frac{25}{32}$	$\frac{26}{32}$	$\frac{27}{32}$	$\frac{28}{32}$	$\frac{29}{32}$	$\frac{30}{32}$	$\frac{31}{32}$	$\frac{32}{32}$

Using the fraction chart above, determine the answers to the fraction problems on the left. Then find and circle the complete spelled-out answers in the word search. Hyphens are not included in the word search words.

1. $\frac{1}{2} = \frac{\quad}{4}$
2. $\frac{1}{2} = \frac{\quad}{8}$
3. $\frac{1}{2} = \frac{\quad}{16}$
4. $\frac{1}{2} = \frac{\quad}{32}$
5. $\frac{1}{4} = \frac{\quad}{8}$
6. $\frac{1}{4} = \frac{\quad}{16}$
7. $\frac{1}{4} = \frac{\quad}{32}$
8. $\frac{1}{8} = \frac{\quad}{16}$
9. $\frac{3}{4} = \frac{\quad}{8}$
10. $\frac{3}{4} = \frac{\quad}{16}$
11. $\frac{3}{4} = \frac{\quad}{32}$
12. $\frac{3}{8} = \frac{\quad}{16}$
13. $\frac{5}{8} = \frac{\quad}{16}$
14. $\frac{2}{4} = \frac{\quad}{2}$
15. $\frac{2}{32} = \frac{\quad}{16}$

```
S H T R U O F O W T D Q F L A H E N O J Z D J P
H B H I Q S H O W W T E N S I X T E E N T H S R
X S E W F D E S H T H G I E O W T B A V K U V E
V Z K B O N E T W J D O T I N F G Q L D Z S R R
V P T X U O Z S L O S M X P R O Y K L C T H F L
C A C G R C Q O B Q H X P K X U O H Q N W T N S
Z P G F E E I N M H T J N S R R Y G D G B N V D
B Y C A I S Q E T S N E L F C S X U S J W E M N
Z I Z X G Y S S Z I E H A Y Y I J C B O S E X O
G W O U H T F I S X E W P W K X S B V S H T H C
J Y N G T R R X L S T Z D B K T R G K Y T X P E
V S L E H I H T Y I X U F G C E O M O D N I W S
S G C V S H N E G X I F E Q R E Q Z W O E S N Y
J R B C X T R E C T S B I T Q N W I D U E O R T
N X M E O R B N B E T D K V W T L E I L T W T R
L E U T N U D T B E H Y L X D H V L Q X X T Z I
M Z M U R O G H A N G M Z E T S Q W I E I S Y H
S O A S I F J C W T I N Y K P K H A L H S T S T
S P H F W Y W N Z H E N T F N O H K B D E K I T
N N C I D T H B K S Y I U S W W J S B X V P P H
E O V Z B N H S L R M A Q U D Z K E D Q L E B G
S I X T E E N T H I R T Y S E C O N D S E X E I
Q N P N S W M M X G S Z R M S Y K T A J W O J E
E K O T Y T H M S H T H G I E X I S J T T B W I
```

Name ______________________________ Date ________________

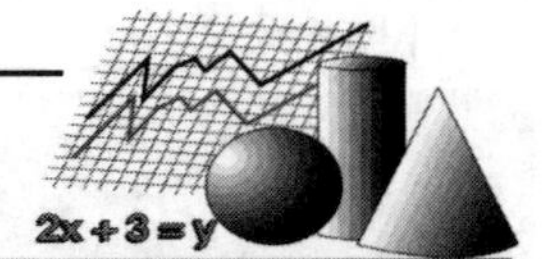

Decimals

Decimals are based on the number ten. Going from right to left, every place is ten times larger than that of the place to the right. A **decimal point** separates whole numbers on the left from those numbers that are less than one on the right.

Example: 555.5

The rightmost 5 is equal to five tenths.
The next 5 is equal to five ones, which is 10 x 0.5.
The next 5 is equal to five tens, which is 10 x 5.
The leftmost 5 is equal to five hundreds, which is 10 x 50.

The number above is pronounced "five hundred fifty-five and five tenths." The decimal point is always pronounced as "and."

When decimals are added or subtracted, they must be lined up vertically according to the decimal points.

Examples:

$$\begin{array}{r} 0.06 \\ 1.783 \\ +\ 942.5 \\ \hline 944.343 \end{array} \qquad \begin{array}{r} 16073.5 \\ -\ 891.043 \\ \hline 15182.457 \end{array} \qquad \begin{array}{r} 3256.32 \\ -\ 733.1225 \\ \hline 2523.1975 \end{array}$$

Quiz

Write the number on the blank for each statement. Remember the decimal point.

1. One and six tenths ____________
2. Eight hundredths ____________
3. Seven and one tenth ____________
4. Seven and ten hundredths ____________
5. One hundred seven and sixty-seven hundredths ____________
6. Four hundred thirty-two and seven thousandths ____________
7. Fifty-six thousandths ____________
8. One thousand seventy-three and four tenths ____________
9. Three thousandths ____________
10. Three and ninety-nine hundredths ____________

Name ______________________ Date ________________

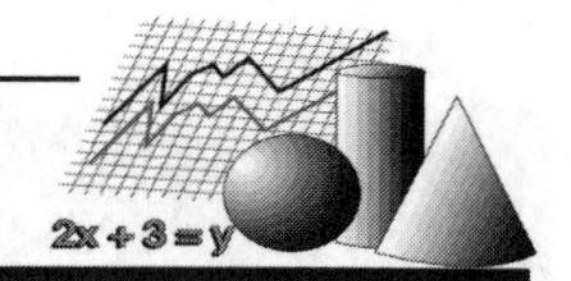

Decimals Hidden Number Puzzle

Solve the following problems involving decimal numbers and then place the answers on the blanks at the right. Place the decimal point on one blank in each answer. When you are finished, the numbers in the circled blanks will reveal the speed of light in statute miles per second.

1. 682.9 + 456.2 = ___ (___) ___ ___ ___ ___
2. 77.06 - 10.1 = ___ ___ ___ ___ ___
3. 302.54 - 121.95 = ___ (___) ___ ___ ___ ___
4. 49.66 - 25.321 = ___ ___ ___ ___ ___ ___
5. 1.482 + 0.387 = ___ ___ ___ (___) ___
6. 51.6 + 0.074 = ___ ___ ___ ___ ___ ___
7. 2.0135 - 0.967 = ___ ___ ___ ___ ___ ___
8. 0.00687 + 64.29 = ___ ___ ___ (___) ___ ___ ___ ___
9. 1.236 + 5.477 = ___ ___ ___ ___ ___
10. 9.975 - 9.156 = ___ ___ (___) ___ ___
11. 323.647 - 7.0922 = ___ ___ ___ ___ ___ ___ ___ ___
12. 4.0126 - 1.80095 = ___ ___ ___ ___ ___ ___ ___
13. 0.13 + 17.3824 = ___ ___ ___ ___ ___ (___) ___
14. 10,633.2 + 5,409.7 = ___ ___ ___ ___ ___ ___ ___
15. 12.11 - 6.33 = ___ ___ ___ ___
16. 35,108.9 - 13,009.3 = ___ ___ ___ ___ ___ (___) ___
17. 57.14 - 25.08 = ___ ___ ___ ___ ___
18. 3.619 + 0.482 = (___) ___ ___ ___ ___
19. 229.25 + 86.982 = ___ ___ ___ ___ ___ ___ ___
20. 197.2 - 73.46 = ___ ___ ___ ___ ___ ___

Hidden Number: The speed of light is ________________ statute miles per second.

Name ______________________ Date ______________

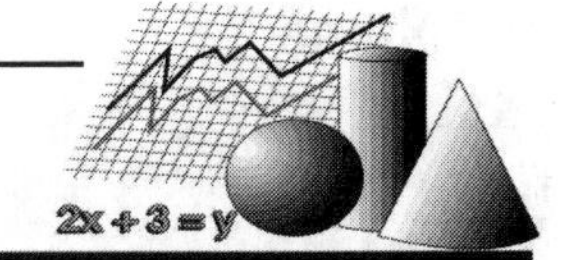

Decimals Crossword Puzzle

You may be most familiar with decimal numbers used to represent amounts of money. Complete the problems below, and then write the answers as decimal numbers with dollar signs in the crossword puzzle.

ACROSS

1. The Jacksons' new car cost $28,682.45. The license fee was $78.50, and sales tax was $1,720.95. Insurance for a year was $913.06. What was the total spent for the new car?
2. Rico, Luis, and Rita were given a large jar of change by their grandparents. If there is $51.54 in the jar, how much will each child get when the money is split evenly?
4. Miguel had $79.32 in his checking account. If he bought a new baseball glove for $57.98, how much money does he have left?
5. Sherry took $48.00 to the mall. She bought a shirt for $17.98, a pair of shorts for $15.75, and a purse for $12.77. How much money did she have left?
7. Lu got money from four people for her birthday. She received $5.00, $8.50, $2.75, and $15.00. What is the total?
8. Maria has $7,095.00 in her savings account. If she has deposited the same amount every year for 11 years, how much does she deposit each year?
9. Frank plans to attend college where the tuition is $14,775.00 a year. How much will tuition for four years cost if the price does not increase?

ACROSS (continued)

10. The Wongs took out a loan for $175,000.00. If they used $148,392.60 to purchase a new house, how much did they have left over to buy furnishings for the house?

DOWN

1. If electricity for the Roberts' house costs $14.09 a week, how much do the Roberts spend on electricity for eight weeks?
3. Stacy makes $12.50 for each yard she mows. How much money will she have after mowing nine yards?
4. Jason gets an allowance of $5.50 a week. How much will he have after three weeks?
5. Sarah earned $16.50, $9.75, and $10.00 by babysitting. She then bought a new computer game for $28.99. How much money does she have left?
6. Brenda received the following tips during the week at her after-school job: $9.75, $17.33, $6.80, $12.14, $14.42. What were her total tips for the week?
7. Kimiko made $56.00 by selling her seven kittens. What was the price of each kitten?
8. The Burger Palace charged Leo's parents $71.50 for his birthday party. There were 22 kids at the party. What was the cost for each child?

Name ______________________________ Date ____________________

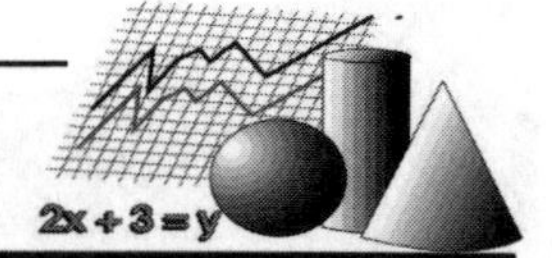

Decimals Number Search

6	4	9	.	9	1	3	5	3	0	2	9	1	4	2	4	2	0
9	5	2	7	5	4	7	2	1	4	5	8	9	0	1	7	4	3
1	.	3	1	.	7	0	6	4	0	9	0	6	3	5	1	0	9
1	4	9	0	4	5	6	.	2	7	2	.	1	2	4	1	2	4
.	6	4	.	1	9	8	3	0	1	.	2	2	8	1	9	.	8
9	1	.	9	8	3	1	9	4	0	3	0	4	7	1	3	4	6
0	8	6	7	2	9	5	3	7	8	.	8	5	4	0	2	6	5
8	1	3	5	3	2	0	5	9	3	6	2	2	0	1	8	.	7
2	7	4	9	0	7	2	3	.	8	7	1	4	0	.	0	7	8
6	2	2	0	3	5	3	0	2	9	1	2	.	5	4	2	6	.
5	3	0	8	9	3	.	4	3	7	5	2	6	3	7	4	9	7
4	3	5	4	7	5	7	1	6	.	9	2	8	4	2	0	0	1
0	5	9	3	.	6	0	5	4	0	4	3	6	.	7	3	.	4
.	4	4	2	8	1	7	3	6	0	4	3	6	2	9	5	3	7
0	2	8	6	7	2	1	2	0	6	3	4	7	.	6	3	6	.
2	1	.	.	4	.	9	6	5	.	0	1	2	9	9	1	4	0
8	6	4	8	2	1	5	7	4	6	2	7	1	3	5	.	2	1
.	3	0	9	9	.	0	9	6	7	9	8	8	.	7	6	2	7

Solve the problems below involving decimals. Then find and circle the answers to each problem in the number search above. Numbers may be horizontal, vertical, diagonal, backwards, or forwards.

1. 20.18 - 9.21 = ____________
2. 5.5 ÷ 1.6 = ____________
3. 567.034 + 413.174 = ____________
4. 90.123 x 0.11 = ____________
5. 12.4 x 3.1 = ____________
6. 5.7 + 3.1 = ____________
7. 0.324 ÷ 7.2 = ____________
8. 109.2 x 22 = ____________
9. 8600 ÷ 4.3 = ____________
10. 67.005 + 7.096 = ____________
11. 4.67 - 3.45 = ____________
12. 907.12 - 98.001 = ____________
13. 21.76 ÷ 3.2 = ____________
14. 8.32 + 2.18 = ____________
15. 89 x 0.2 = ____________
16. 56.001 - 47.067 = ____________
17. 12.11 x 2.1 = ____________
18. 45.78 + 32.54 = ____________
19. 7.84 ÷ 5.6 = ____________
20. 8977.1 - 988.4 = ____________

Name ______________________________ Date ____________________

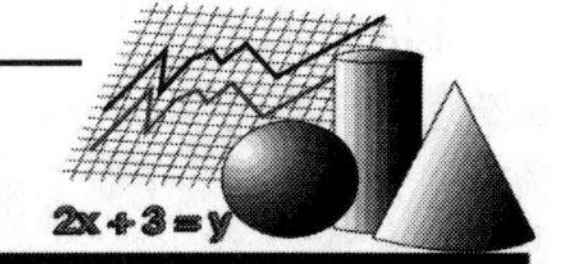

Percentages

The word **percent** means "per hundred." For example, 30 percent means "30 in every hundred." Like decimals and fractions, a percentage is a part of a whole.

A percent sign is used to show that a number is a percent.

Examples: 50% 10% 95% 20.5%

A percent can be changed to a decimal so that it can be used in a mathematical computation by dividing the percent by 100.

Examples: 25% = 25 ÷ 100 = 0.25 42.7% = 42.7 ÷ 100 = 0.427

Another way of converting a percent to a decimal is to move the decimal point two places to the left. (This is what happens when a number is divided by 100.)

Examples: 50% = 0.50 20.5% = 0.205 7% = 0.07

To find a certain percent of a number, convert the percent to a decimal and then multiply it by the number.

Example: What is 28% of 650?
28% = 0.28 0.28 x 650 = 182

Quiz

Change each of the following decimals to a number with a percent sign.

1. 0.40 = __________ 2. 0.67 = __________ 3. 0.89 = __________
4. 2.32 = __________ 5. 0.3429 = __________ 6. 0.09 = __________
7. 3.03 = __________ 8. 6.54 = __________ 9. 0.008 = __________

Change the following percents to decimals.

10. 15% = __________ 11. 0.4% = __________ 12. 512% = __________
13. 0.31% = __________ 14. 1765% = __________ 15. 2% = __________
16. 100% = __________ 17. 0.287% = __________ 18. 78% = __________

Figure out the percentage of each number given.

19. What is 53% of 225? __________ 20. What is 0.5% of 1,000? __________
21. What is 28.75% of 375? __________ 22. What is 78.75% of 980? __________
23. What is 350% of 250? __________ 24. What is 82% of 500? __________

Name ______________________________ Date __________________

Percentages Hidden Picture Puzzle

79.8	179	378	5.672	45	983	2074	211	90	6501	63.22
425	1.275	0.25	1.06	12.98	616	7053	0.418	1.98	0.8	107
35	16.25	8.36	512	7.32	24.09	532	906	144	171.92	92
135	17.62	7.54	139	90.3	267	5223	431	1.95	82.08	11.567
569	293	3602	4576	163.02	10.62	26.25	1.86	3.554	6.22	6.8
12.34	61.52	487	6.35	156.4	361	16.038	728	1.342	39.5	34
46	82.67	287.3	14.12	790	18.63	6054	1.683	952	10.87	105.1
34.90	726	147	8.514	5.832	67.20	2.39	12.15	10.15	521	82.6
799	1462	555	14.09	0.0665	1.2	4.5	195.2	898	633.2	9653
385	4.754	78.2	643	153.9	0.098	76.1	383.6	62	14.13	80.2
16548	909	7.665	1.8325	7.85	58	311	6.9	0.364	28.9	0.4

Solve the problems below. Then find each answer in the grid above and shade the square in which the answer appears to reveal a hidden picture.

1. 20% of 782 = ____________
2. 5% of 90 = ____________
3. 0.8% of 100 = ____________
4. 75% of 35 = ____________
5. 9.5% of 88 = ____________
6. 125% of 13 = ____________
7. 0.04% of 245 = ____________
8. 38% of 950 = ____________
9. 200% of 72 = ____________
10. 1.8% of 675 = ____________
11. 0.5% of 50 = ____________
12. 118% of 9 = ____________
13. 47.3% of 18 = ____________
14. 56% of 307 = ____________
15. 16.2% of 99 = ____________
16. 4% of 30 = ____________
17. 15% of 8.5 = ____________
18. 247% of 66 = ____________
19. 0.95% of 7 = ____________
20. 22% of 9 = ____________

Name ______________________________ Date ________________

Percentages Crossword Puzzle

Complete each problem below and then place the answer in the crossword puzzle. Include decimal points and percent signs in the answers in the crossword puzzle.

ACROSS

2. Convert 6.5% to a decimal. ________
4. Convert 77% to a decimal. ________
6. Convert 0.00523% to a decimal. ________
8. Convert 42.303 to a percent. ________
9. Convert 0.1695 to a percent. ________
11. Convert 0.01 to a percent. ________
15. Convert 579.51% to a decimal. ________
16. Convert 32.064 to a percent. ________
18. Convert 5.0069 to a percent. ________
19. Convert 6.895 to a percent. ________

DOWN

1. Convert 892.1% to a decimal. ________
2. Convert 0.0087 to a percent. ________
3. Convert 0.000719 to a percent. ________
5. Convert 28.1054% to a decimal. ________
7. Convert 191% to a decimal. ________
10. Convert 0.09542 to a percent. ________
12. Convert 0.00102 to a percent. ________
13. Convert 1,000.75% to a decimal. ________
14. Convert 12.3635% to a decimal. ________
17. Convert 70.0341% to a decimal. ________

Name ______________________________ Date ________________

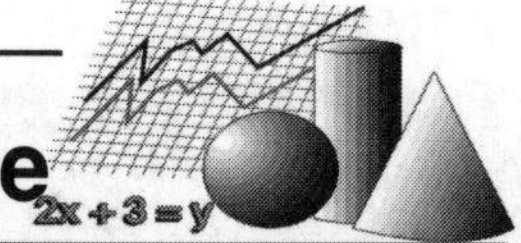

Percentages Connect-the-Dots Puzzle

Complete the following selection and problems using the terms and numbers in the puzzle picture below. Then fill in the circle next to each correct answer. Once you are finished with all the questions, connect all the dots, starting with the answer to #1 and proceeding through #12c, to find a hidden picture.

A decimal is a number that may also be written in (1) __________ form or percent form. The decimal 0.73 could be written as the fraction (2) ________ or as (3) ________ %. When changing a fraction to a decimal, the (4) ______________ must be divided by the (5) ______________. The decimal from this division can be converted to a percent by moving the decimal (6) __________ (7) __________ places to the (8) __________. To change 0.73 to a (9) __________, the (10) __________ point is moved two places to the right, and the number is then written as 73%.

11. Sally can buy a pair of jogging shoes for $98. She can buy the same shoes across town for 15% less. However, it will cost her $10 to get across town.
    a) What will be the price of the shoes across town? __________
    b) How much cheaper are the shoes across town? __________
    c) Are the shoes still cheaper if Sally includes the cost of traveling across town? __________

12. Esteban can buy a shirt for $34. If he buys a second shirt, the price of the second shirt will be reduced 35%.
    a) How much does the second shirt cost? __________
    b) How much cheaper is the second shirt? __________
    c) The total cost of both shirts is __________.

fraction ○ ○ denominator ○ point ○ two ○ problem ○

$77.35 ○

○ $\frac{73}{100}$ ○ numerator left ○

$9.60 ○ ○ right

number ○ ○ 73 six ○

○ no

division ○ $19.80 ○

percent ○ four ○ $11.90 ○

○ ratio

$27.25 ○ ○ $22.10

$49.60 ○ $56.10 ○

○ $\frac{100}{73}$ decimal ○ $83.30 ○ $14.70 ○ ○ yes ○ 0.73

Name ______________________________ Date ________________

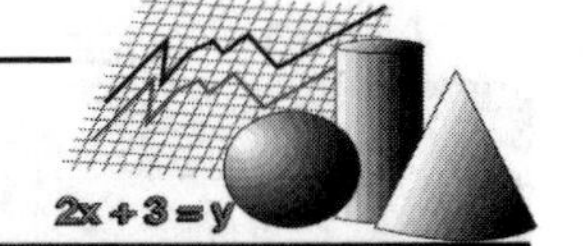

Geometry: Angles

Geometry is the branch of mathematics that deals with points, lines, planes, and figures and their properties, measurements, and mutual relations in space.

An **angle** is the shape made by two straight lines, line segments, or rays meeting at a common point called the **vertex**. The correct way to write an angle is to write "angle ABC," "∠ABC," or "∠B." The middle letter is always the vertex. If you use only one letter, it must be the vertex.

Angles are measured in degrees. To measure angles you need to have a protractor. Angles that measure less than 90° are called **acute angles** (∠DEF in the diagram below). Angles that measure 90° exactly are called **right angles** (∠CBA below). Angles that measure more than 90° are called **obtuse angles** (∠CED below).

If the sum of two angles equals 90°, they are called **complementary angles** (∠GHI and ∠IHJ below). If the sum of two angles equals 180°, they are called **supplementary angles** (∠GHI and ∠IHK below). The angle measure of a line is always equal to 180°. When two angles have the same vertex and a side in common, they are called **adjacent angles** (∠AHD and ∠DHK below).

If two lines or line segments intersect, the angles formed are called **vertical angles**. Opposite vertical angles are **congruent** or equal. In the diagram below, ∠DEF and ∠CEI are opposite vertical angles, and ∠CED and ∠IEF are opposite vertical angles.

When two parallel lines are crossed by a third line, special angle relationships exist. First, the line that crosses the two parallel lines is known as the **transversal**. In the diagram below, lines x and y are parallel, and line t is the transversal. There are eight angles created by the intersection of these lines.

Alternate exterior angles are the angles that lie on the outside of the parallel lines and on opposite sides of the transversal. Alternate exterior angles are congruent or equal. So ∠CED and ∠IHK are congruent and ∠DEF and ∠GHI are congruent.

Alternate interior angles are the angles that lie on the inside of the parallel lines and on the opposite sides of the transversal. Alternate interior angles are congruent. So ∠CEI and ∠DHK are congruent and ∠IEF and ∠GHD are congruent.

Corresponding angles are the angles that lie on the same side of the transversal and either both lie above or below the parallel lines. Corresponding angles are congruent. So ∠DEF and ∠DHK are congruent, ∠IEF and ∠IHK are congruent, ∠CED and ∠GHD are congruent, and ∠CEI and ∠GHI are congruent.

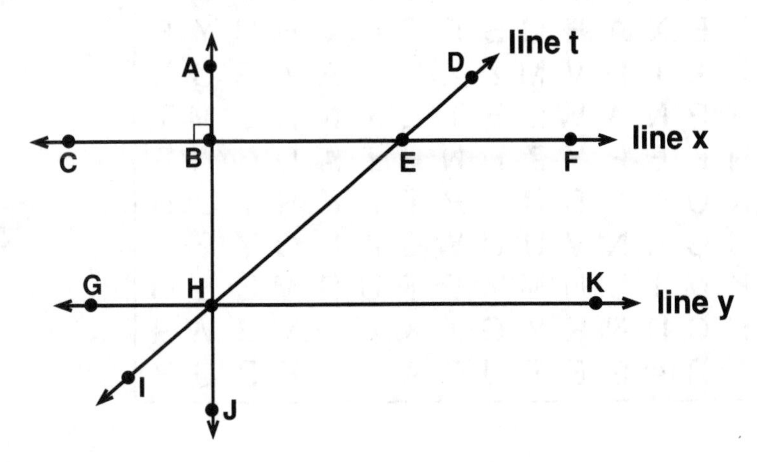

Name ______________________ Date ______________

Angles Word Search

List the names for these angles and then find them in the word search below.

1. ∠ADB ______________
2. ∠AHI ______________
3. ∠FGB ______________
4. ∠FGB and ∠BGI ______________
5. ∠JDK and ∠KDC ______________
6. ∠BDE and ∠BGI ______________
7. ∠ADB and ∠KDJ ______________
8. ∠CDK and ∠BGI ______________
9. ∠CDA and ∠ADB ______________
10. ∠KGI and ∠CDB ______________

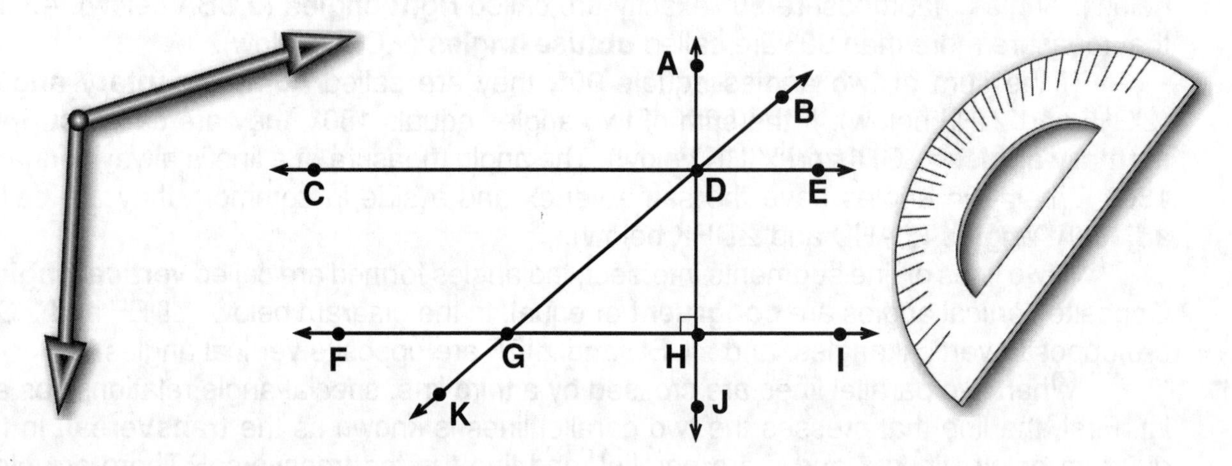

M A R S C O R R E S P O N D I N G M D
V N O M P O J T E S U W X A C I D B A
F G I T D R Y L A D J C G Z I T B F L
E C R U C F F V A F L R I G H T I W T
S R E B W O I L V C I V B G T X I N E
U R T S T K M L X D I I W I Z X W C R
T K N F U V Z P G F F T L V R J Q C N
B P I S X P D L L L V B R T T H J E A
O S E C K Z P M A E U V C E Q G Q S T
M T T R I N P L W H M T E T V D S C E
G T A Z I B X A E G B E F C V H U Y E
I I N H Z X T N V M Z X N I A Z R C X
X X R N A P N A W L E T E T N R D M T
O R E L N L E P Y Z H N R Z A D O I E
X K T M K U C Y D R S P T Z E R T L R
O B L C U C A N V U U W B A L Q Y F I
D B A Z K M J L N M Y E T U R M B L O
K T B O I C D N K Y G J X J Z Y B W R
H W O H E R A G E T U C A T F F P U P

Word List

vertical
complementary
adjacent
corresponding
acute
obtuse
alternate interior
supplementary
right
alternate exterior

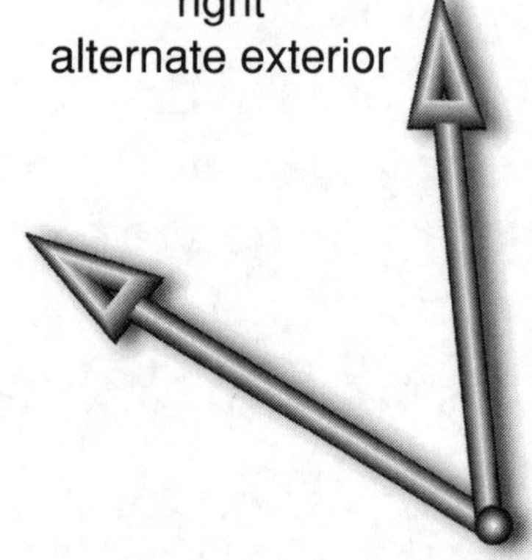

Name ______________________________ Date ______________

Geometric Figures

Geometry is the branch of mathematics that deals with points, lines, planes, and figures and their properties, measurements, and mutual relations in space.

Two-dimensional figures are named according to how many sides they have. A three-sided figure is a **triangle**, a four-sided figure is a **quadrilateral**, and so on. A figure is called **regular** when each side is of equal length and each angle is of equal measure. For a triangle this makes a regular triangle an equilateral triangle. The regular quadrilateral is the square.

The total **sum of all angles** in a triangle is exactly 180°. If the triangle is regular, this means that each angle is exactly 60°.

To determine the total sum of the angles for figures with more than three sides, we must first determine how many diagonals the figure has. A **diagonal** is a line drawn from one vertex (corner) of a figure to another that cuts a figure into triangles. The square at the right has only one diagonal. This diagonal cuts the square into two separate triangles. Since each triangle is equal to 180°, then the total sum of a square's angles is eactly 2 x 180° or 360°. Since the square has four angles, each angle is 360° ÷ 4 = 90°. This process can be completed for all two-dimensional figures.

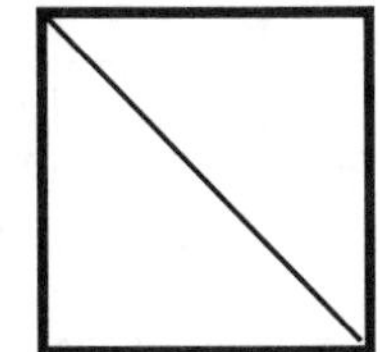

Quiz

Fill in the table for each regular figure. You may need to draw each figure on your own paper and determine how many diagonals the figure has.

Name of figure	**Number of sides**	**Number of angles**	**Number of diagonals**	**Number of triangles**	**Angle sum of one triangle**	**Angle sum of the figure**	**Measure of each angle**
Triangle	3	3	0	1	180°	180°	60°
Quadrilateral	4	4	1	2	180°	360°	90°
Pentagon	5						
Hexagon	6						
Heptagon	7						
Octagon	8						
Decagon	10						
Dodecagon	12						

Name ______________________ Date ______________

Geometric Figures Crossword Puzzle

Complete the crossword puzzle using the clues given below. The clues show two- and three-dimensional geometric figures. Fill in their names in the appropriate places. You may need to use a math textbook, dictionary, or other source to find the correct names.

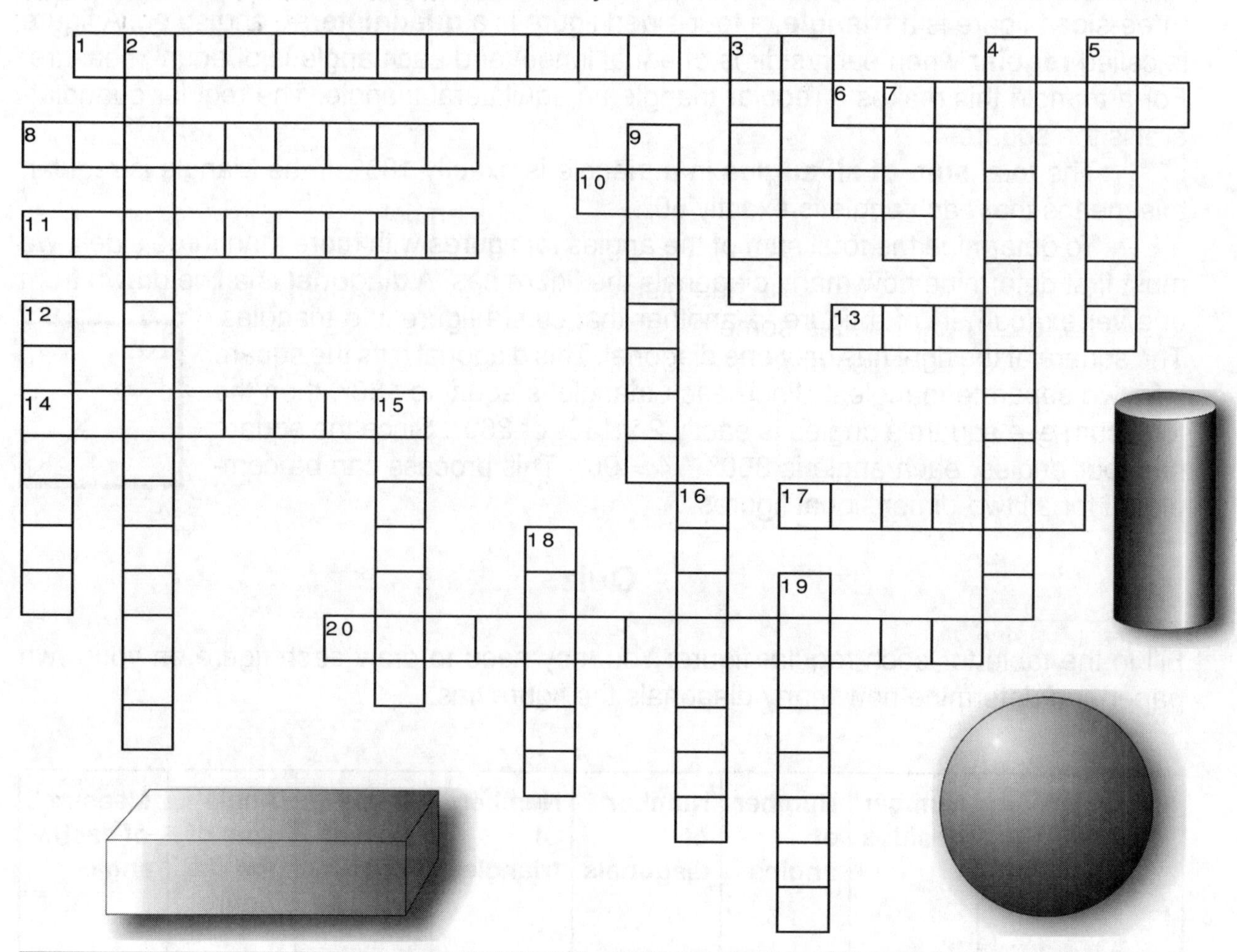

ACROSS

1.
6.
8.
10.
11.
13.
14.
17.
20.

DOWN

2.
3.
4.
5.
7.
9.
12.
15.
16.
18.
19.

Name ______________________________ Date ________________

Algebra: Algebraic Equations

Algebra is a mathematical system using symbols to generalize certain arithmetical operations and relationships.

An **equation** is a statement of equality between two quantities joined by an equal sign (=). This means that each number, expression, or variable on either side of the equal sign represents the same value.

Examples: $16 = 16$ $2 + 5 = 5 + 2$ $18 = 6 \cdot 3$ $4 + 3 = 9 - 2$ $2n = 12$

A **variable** is a letter or symbol that represents an unknown number in an expression. For example, in the equation $2n = 12$, the n is a variable that stands for an unknown number. We can determine what the n stands for, because we know that in order for the equation to be true, 2 times some number must equal 12. What number taken times 2 equals 12? We could also divide 12 by 2 to get the answer. The variable n stands for the number 6.

There are other parts of an algebraic expression. The number that multiplies a variable is called the **coefficient**. In $2n = 12$, the 2 is the coefficient. A number that is added to or subtracted from a variable is called the **constant**. In $2n + 2 = 12$, the constant is +2.

Quiz

Identify the coefficient(s), variable(s), and constant(s) in each expression below.

	Coefficient(s)	Variable(s)	Constant(s)
1. $7x + 3$	________	________	________
2. $\frac{1}{4}n - 9$	________	________	________
3. $12a + 6b - 4c + 2$	________	________	________
4. $11 + 3y$	________	________	________
5. $\frac{2}{5}g + 8m - 10$	________	________	________

Solve the following equations for the value given for each variable.

6. $(b = 4)$ $8b + 3 =$ ____________
7. $(n = 2)$ $\frac{18}{n} - 5 =$ ____________
8. $(a = 5)$ $11 - 6a =$ ____________
9. $(c = -10)$ $-2c + 9 =$ ____________
10. $(s = 4, t = 0)$ $7s + 5t - 13 =$ ____________

Name ______________________ Date ______________

Algebraic Equations Crossword Puzzle

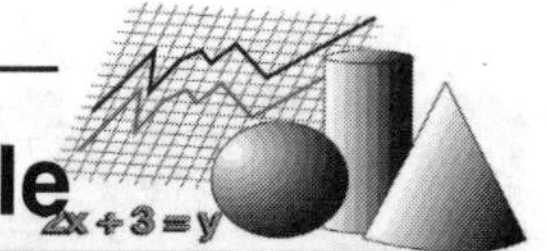

Write the equation for each word problem below. Then solve each equation for the variable and write the answer as a word in the crossword puzzle.

Show your algebra steps neatly on notebook paper!

30 - x = 12

2x + 3 = 11

ACROSS

1. Twenty-five divided by an unknown number plus three equals eight. ______________
2. Nine plus twenty divided by an unknown number equals eleven. ______________
4. Three times an unknown number plus two equals eleven. ______________
6. Negative six plus an unknown number divided by two equals five. ______________
9. Nine times an unknown number minus fourteen equals sixty-seven. ______________
11. Eighteen minus two times an unknown number equals fourteen. ______________
12. Twenty-one divided by an unknown number plus ten equals thirteen. ______________
13. An unknown number plus twelve equals twenty-eight. ______________

DOWN

1. Seventy-five divided by an unknown number equals five. ______________
3. Six times an unknown number minus three equals forty-five. ______________
5. Five times an unknown number minus ten equals fifty. ______________
7. Seven plus four times an unknown number equals fifty-one. ______________
8. Two times an unknown number plus seven equals sixty-seven. ______________
10. Nine plus seven times an unknown number equals thirty-seven. ______________
12. Sixty divided by an unknown number plus five equals fifteen. ______________

Name ______________________________ Date ________________

Algebra: Rectangular Coordinate Systems

A **rectangular coordinate system** is developed by drawing two perpendicular lines that are then numbered like a double number line from the point of origin. The vertical line is called the ***y* axis**, and the horizontal line is called the ***x* axis**. The **origin**, or reference point, is zero. (See diagram below.)

The rectangular coordinate system can be used to locate points in a plane. A coordinate point is located by its distance from the *x* and *y* axes. Each point has two numbers locating it. These numbers are called **coordinates**. The numbers are written inside parentheses and are separated by a comma. The *x* axis number is written first, and the *y* axis number is written second.

Example: The coordinates for Point A are (3, 5).
This means the point is three places to the right of the *y* axis and five places above the *x* axis.
To locate this point, move three places to the right from the origin along the *x* axis, and then move up five places.

Points with negative numbers as coordinates are found by moving to the left of the *y* axis and below the *x* axis.

Example: The coordinates for Point B are (-2, -6).
To locate this point, move two places to the left from the origin along the *x* axis, and then move down six places.

Coordinates may consist of both positive and negative numbers. For example, Point C is located at (1, -3) and Point D is located at (-4, 2).

Quiz

Locate each point on the rectangular coordinate system at the right. Place a dot and the letter at the location of each point.

1. Point E (1, 5)
2. Point F (-3, 4)
3. Point G (-5, -2)
4. Point H (4, 0)
5. Point I (0, 6)

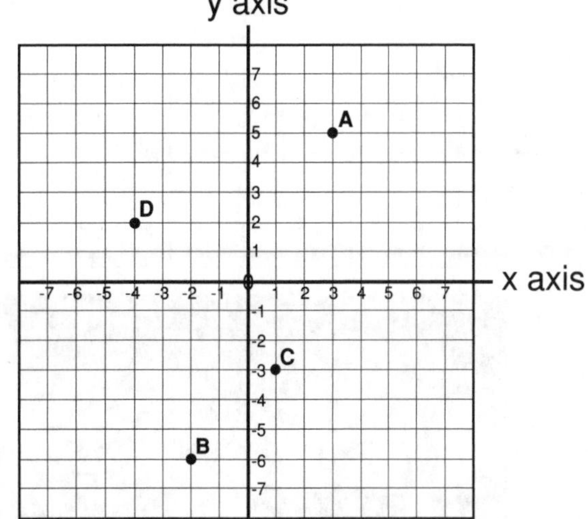

Name ______________________________ Date __________________

Rectangular Coordinate Systems Puzzle

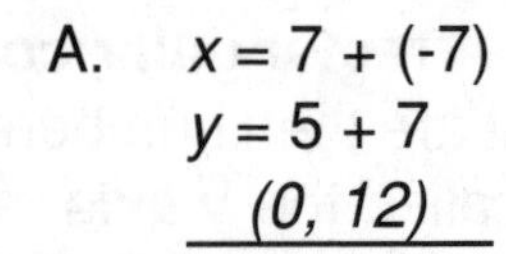

For each set of coordinates there is exactly one point located on the rectangular coordinate system. Locating a point is called plotting the point. Use this rectangular coordinate system below to locate the points on the right. Place a dot on the graph to locate each point. Then place the letter representing each point next to each dot. Solve each equation on the right to come up with the coordinates for each point. Remember, the first number in a coordinate pair is the x coordinate and the second number is the y coordinate. The first one is completed for you. When all the points are plotted, connect the dots in order from A to J and back to A to reveal a picture.

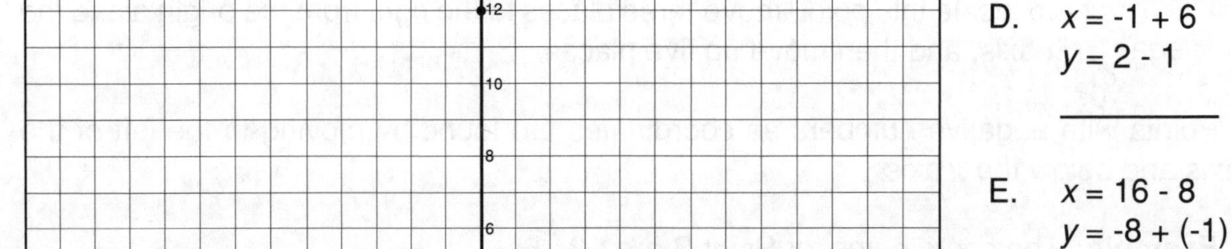

A. $x = 7 + (-7)$
$y = 5 + 7$
(0, 12)

B. $x = -10 + 13$
$y = 15 - 8$
____

C. $x = 6 + 4$
$y = 10 + (-3)$
____

D. $x = -1 + 6$
$y = 2 - 1$
____

E. $x = 16 - 8$
$y = -8 + (-1)$
____

F. $x = 8 - 8$
$y = -5 + 5$
____

G. $x = -8 + 0$
$y = -11 + 2$
____

H. $x = -10 - (-5)$
$y = -8 + 9$
____

I. $x = -20 + 10$
$y = 4 + 3$
____

J. $x = -7 - (-4)$
$y = 14 + (-7)$
____

Name ______________________ Date ______________

Metrics

The symbol for a metric measurement has two parts at most. The **stem** consists of the actual unit name of the type of measurement used. The **prefix** comes before the stem, if it is used, and describes how large the unit actually is.

These are the stems or units for each type of metric measurement.

Type of Measurement	Unit Name	Symbol	U.S. Standard Equivalent
Length	meter	m	1 meter = 1.0936 yards
Mass/Weight	gram	g	1 gram = 0.03527 ounces
Volume/Capacity	liter	L or l	1 liter (fluid) = 1.057 quarts

These are the most common metric prefixes listed from largest to smallest.

Prefix Name	Symbol	Decimal Value	Fraction Value	Description
giga	G	1,000,000,000		billion
mega	M	1,000,000		million
kilo	k	1,000		thousand
hecto	h	100		hundred
deca (or deka)	da	10		ten
deci	d	0.1	1/10	tenth
centi	c	0.01	1/100	hundredth
milli	m	0.001	1/1,000	thousandth
micro	µ	0.000 001	1/1,000,000	millionth
nano	n	0.000 000 001	1/1,000,000,000	billionth

Quiz

Circle the most reasonable metric measure for each item listed.

1.	Distance from home to school	a) 10 m	b) 10 km	c) 10 cm
2.	Weight of a lunch box	a) 980 mg	b) 980 kg	c) 980 g
3.	Volume of liquid in a small bottle of juice	a) 350 L	b) 350 mL	c) 350 kL
4.	Width of a tablet of paper	a) 21 cm	b) 0.5 m	c) 82 mm
5.	Weight of a math book	a) 6 kg	b) 10.2 kg	c) 1.5 kg
6.	Volume of a large jug of milk	a) 4 L	b) 890 mL	c) 92 cL
7.	Length of an ink pen	a) 25 cm	b) 0.15 m	c) 300 mm
8.	Weight of a candy bar	a) 60 g	b) 1.4 kg	c) 12 mg

Name ______________________________ Date ____________________

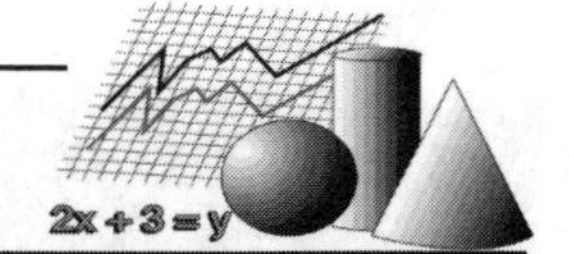

Metrics Crossword Puzzle

Complete the crossword puzzle using the clues given below.

ACROSS

4. Comes before the stem and describes how large the unit actually is
5. Ten grams equals one ______.
6. µ; equals one-millionth
9. Milli (m) means ______.
15. Deci (d) means ______.
16. M; equals one million
18. Basic metric unit for measuring weight
19. Basic metric unit for measuring volume or capacity

DOWN

1. Ten milligrams equals one ______.
2. 1,000 meters equals one ______.
3. 0.1 centimeters equals one ______.
7. Centi (c) means ______.
8. Basic metric unit for measuring length
9. Kilo (k) means ______.
10. Deca (or deka) (da) means ______.
11. Hecto (h) means ______.
12. Consists of the actual unit name of the type of measurement used
13. 10,000 centigrams equals one ______.
14. n; equals one-billionth
17. G; equals one billion

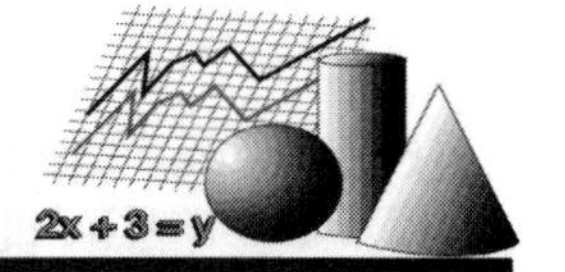

Answer Keys

Place Value Quiz (page 1)

1. 1, 4
2. 2, 1, 3, 9, 5
3. 3, 1, 2, 3
4. 1, 8, 9, 0, 8
5. 1, 2, 3
6. 0, 3, 9
7. 7, 6, 5, 4
8. 1, 0, 9, 8, 6
9. 9, 0, 4, 5, 7
10. 9, 4, 5, 8, 3

Place Value Crossword Puzzle (page 2)

Addition Quiz (page 3)

1. 76
2. 8.9
3. 62
4. 45.9
5. 701
6. 58.50
7. 2,485
8. 751.79
9. 6,255

Addition Magic Squares (page 4)

6	(11)	4
5	7	(9)
(10)	3	8

Magic Number 21

7	26	(21)
(32)	18	(4)
15	10	29

Magic Number 54

64	(74)	141
(170)	93	16
45	(112)	122

Magic Number 279

88	(237)	(143)
(211)	156	101
169	(75)	224

Magic Number 468

(22)	107	90
(141)	73	(5)
56	(39)	124

Magic Number 219

(63)	5	(55)
33	(41)	49
(27)	77	19

Magic Number 123

(151)	(60)	134
(98)	115	(132)
96	170	79

Magic Number 345

68	(323)	272
(425)	(221)	17
170	119	(374)

Magic Number 663

	Teacher Check	

Magic Number ______

Addition Word Search (page 5)

1. 107
2. 389
3. 70.6
4. 9,290
5. 1,001
6. 696.2
7. 1,205.4
8. 6,512
9. 18,132
10. 145.72
11. 2,929
12. 115.326

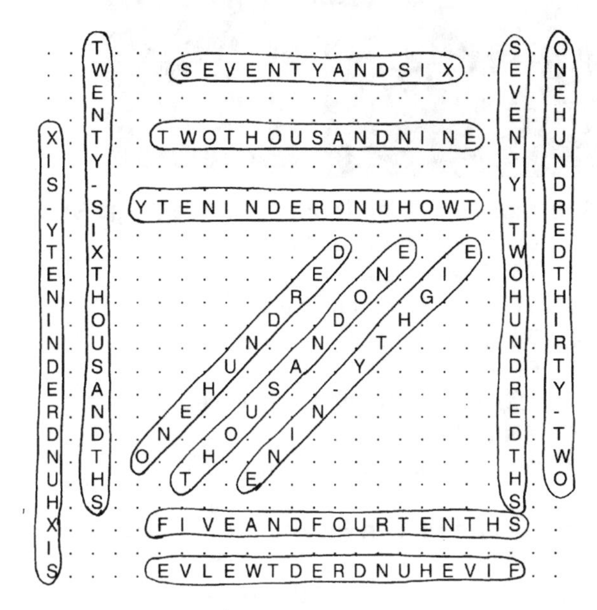

Addition Crossword Puzzle (page 6)

Subtraction Quiz (page 7)

1. 21
2. 3.5
3. 121
4. 27
5. 213
6. 16.06
7. 999
8. 295.97
9. 1,489
10. 1,061.81
11. 20,578
12. 232.392

Subtraction Crossword Puzzle (page 8)

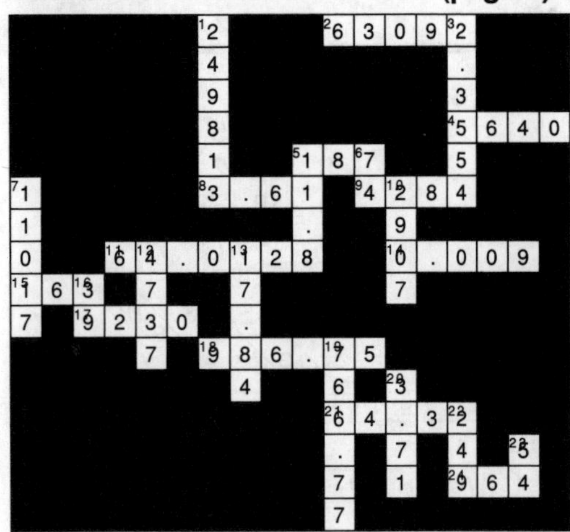

Subtraction Hidden Number Puzzle (page 9)

1. 850; Eight Hundred Fifty
2. 76; Seventy-six
3. 112; One Hundred Twelve
4. 74; Seventy-four
5. 89; Eighty-nine
6. 125; One Hundred Twenty-five
7. 16; Sixteen

Hidden Number: Seventy-one

Subtraction Hidden Picture Puzzle (page 10)

1. 4,576
2. 726
3. 5,223
4. 487
5. 2,603
6. 728
7. 431
8. 139
9. 293
10. 109
11. 1.95
12. 14.12
13. 17.62
14. 10.15
15. 1.86
16. 82.67
17. 2.39
18. 61.52
19. 1.683
20. 6.35

80.2	179	378	5.672	45	983	2074	211	90	6501
425	881	7.64	1.06	12.98	616	7053	0.418	2573	194
35	827	952	512	7.32	24.09	532	906	3356	87.42
135	17.62	7.54	139	90.3	267	5223	431	1.95	82.08
569	293	3602	4576	8.6	924	7645	1.86	3.554	6.22
12.34	61.52	487	6.35	9467	3002	88.4	728	1.342	39.5
46	82.67	287.3	14.12	790	18.63	6054	1.683	952	10.87
34.90	726	147	109	5.832	67.20	2.39	2603	10.15	521
799	1462	555	14.09	9.311	8.6	3072	195.2	898	633.2
385	4.754	78.2	643	153.9	9051	76.1	383.6	62	14.13

Multiplication Quiz (page 11)

1. 108
2. 4.32
3. 2,006
4. 945.28
5. 34,335
6. 19,710
7. 308,730
8. 251,256
9. 249.7440

Multiplication Chart (page 12)

X	1	2	3	4	5	6	7	8	9	10	11	12	13	14	15
1	1	2	3	4	5	6	7	8	9	10	11	12	13	14	15
2	2	4	6	8	10	12	14	16	18	20	22	24	26	28	30
3	3	6	9	12	15	18	21	24	27	30	33	36	39	42	45
4	4	8	12	16	20	24	28	32	36	40	44	48	52	56	60
5	5	10	15	20	25	30	35	40	45	50	55	60	65	70	75
6	6	12	18	24	30	36	42	48	54	60	66	72	78	84	90
7	7	14	21	28	35	42	49	56	63	70	77	84	91	98	105
8	8	16	24	32	40	48	56	64	72	80	88	96	104	112	120
9	9	18	27	36	45	54	63	72	81	90	99	108	117	126	135
10	10	20	30	40	50	60	70	80	90	100	110	120	130	140	150
11	11	22	33	44	55	66	77	88	99	110	121	132	143	154	165
12	12	24	36	48	60	72	84	96	108	120	132	144	156	168	180
13	13	26	39	52	65	78	91	104	117	130	143	156	169	182	195
14	14	28	42	56	70	84	98	112	126	140	154	168	182	196	210
15	15	30	45	60	75	90	105	120	135	150	165	180	195	210	225
16	16	32	48	64	80	96	112	128	144	160	176	192	208	224	240
17	17	34	51	68	85	102	119	136	153	170	187	204	221	238	255
18	18	36	54	72	90	108	126	144	162	180	198	216	234	252	270
19	19	38	57	76	95	114	133	152	171	190	209	228	247	266	285
20	20	40	60	80	100	120	140	160	180	200	220	240	260	280	300

Multiplication Crossword Puzzle (page 13)

Multiplication Hidden Number Puzzle (page 14)

1. 408; Four Hundred Eight
2. 110; One Hundred Ten
3. 36; Thirty-six
4. 684; Six Hundred Eighty-four
5. 290; Two Hundred Ninety
6. 3,060; Three Thousand Sixty
7. 612; Six Hundred Twelve

Hidden Number: 412; Four Hundred Twelve

Division Quiz (page 15)

1. 3R4
2. 15R7
3. 8.75
4. 66
5. 24.53
6. 16.25

Division Hidden Picture Puzzle (page 16)

1. 73
2. 19
3. 21
4. 43
5. 107
6. 212
7. 839
8. 707
9. 56
10. 34
11. 689
12. 82
13. 867
14. 44
15. 51
16. 417
17. 17
18. 98
19. 361
20. 63

80	82	378	672	45	983	73	211	90	651
425	881	21	106	298	707	53	41	573	194
35	87	952	417	212	29	52	906	31	84
51	107	361	867	56	19	17	689	195	88
69	293	362	43	44	24	745	186	354	22
123	65	98	66	967	63	84	728	142	39
46	839	287	112	70	183	34	168	952	87
390	726	147	109	58	67	23	263	101	521
99	146	555	14	93	86	372	152	89	633
385	47	72	643	153	951	71	38	62	141

Division Crossword Puzzle (page 17)

Division Word Search (page 18)

1. 26
2. 42
3. 62
4. 89
5. 9
6. 44
7. 95
8. 67
9. 64
10. 49
11. 32
12. 190
13. 5
14. 18
15. 12
16. 3
17. 4
18. 556
19. 20
20. 75

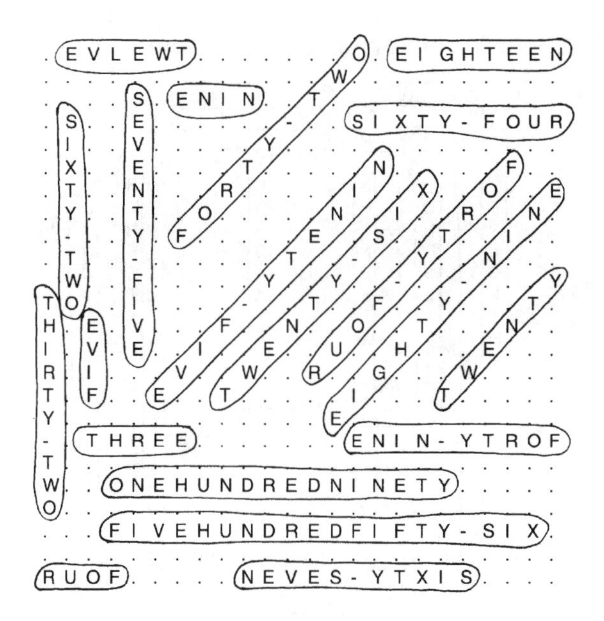

Fractions Quiz (page 20)

1. $\frac{5}{8}$
2. $\frac{9}{8} = 1\frac{1}{8}$
3. $\frac{4}{14} = \frac{2}{7}$
4. $\frac{7}{10}$
5. $\frac{24}{4} = 6$
6. $\frac{10}{56} = \frac{5}{28}$
7. $\frac{493}{54} = 9\frac{7}{54}$
8. $\frac{64}{135}$
9. $\frac{111}{7} = 15\frac{6}{7}$

Fractions Crossword Puzzle (page 21)

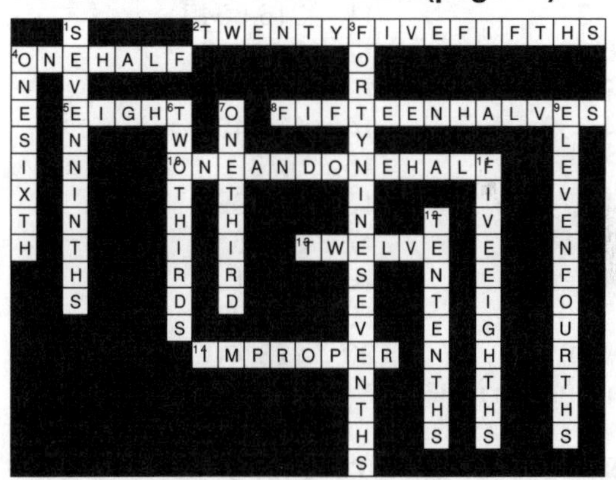

Fractions Word Search (page 22)

1. $\frac{1}{2} = \frac{2}{4}$
2. $\frac{1}{2} = \frac{4}{8}$
3. $\frac{1}{2} = \frac{8}{16}$
4. $\frac{1}{2} = \frac{16}{32}$
5. $\frac{1}{4} = \frac{2}{8}$
6. $\frac{1}{4} = \frac{4}{16}$
7. $\frac{1}{4} = \frac{8}{32}$
8. $\frac{1}{8} = \frac{2}{16}$
9. $\frac{3}{4} = \frac{6}{8}$
10. $\frac{3}{4} = \frac{12}{16}$
11. $\frac{3}{4} = \frac{24}{32}$
12. $\frac{3}{8} = \frac{6}{16}$
13. $\frac{5}{8} = \frac{10}{16}$
14. $\frac{2}{4} = \frac{1}{2}$
15. $\frac{2}{32} = \frac{1}{16}$

Fractions Word Search (page 22) cont.

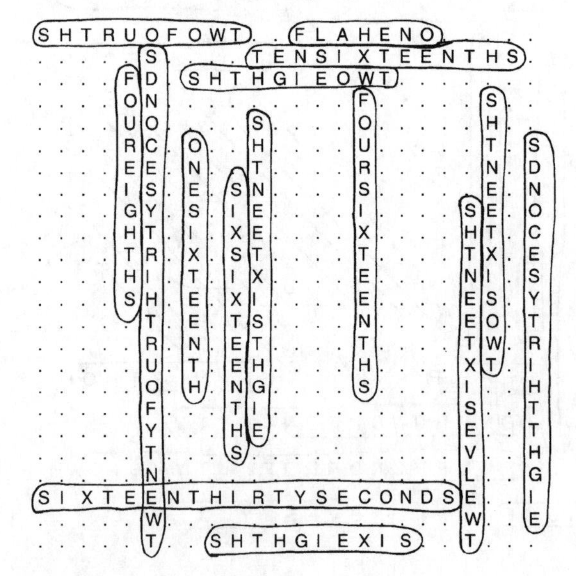

Decimals Quiz (page 23)

1. 1.6
2. 0.08
3. 7.1
4. 7.10
5. 107.67
6. 432.007
7. 0.056
8. 1,073.4
9. 0.003
10. 3.99

Decimals Hidden Number Puzzle (page 24)

1. 1,139.1
2. 66.96
3. 180.59
4. 24.339
5. 1.869
6. 51.674
7. 1.0465
8. 64.29687
9. 6.713
10. 0.819
11. 316.5548
12. 2.21165
13. 17.5124
14. 16,042.9
15. 5.78
16. 22,099.6
17. 32.06
18. 4.101
19. 316.232
20. 123.74

Hidden Number: 186,282.4

Decimals Crossword Puzzle (page 25)

Decimals Number Search (page 26)

1. 10.97
2. 3.4375
3. 980.208
4. 9.91353
5. 38.44
6. 8.8
7. 0.045
8. 2,402.4
9. 2,000
10. 74.101
11. 1.22
12. 809.119
13. 6.8
14. 10.5
15. 17.8
16. 8.934
17. 25.431
18. 78.32
19. 1.4
20. 7,988.7

6 4 9 . 9 1 3 5 3 0 2 9 1 4 2 4 2 0
9 5 2 7 5 4 7 2 1 4 5 8 9 0 1 7 4 3
1 . 3 1 . 4 0 6 4 0 9 0 6 3 5 1 0 9
1 4 6 0 4 5 6 . 8 7 2 . 1 2 4 1 2 4
. 6 4 . 1 9 8 3 0 1 . 2 2 8 1 9 . 8
9 1 . 9 8 3 1 9 4 0 3 0 4 7 1 3 4 6
0 8 6 7 2 9 5 3 7 8 . 8 5 4 0 2 6 5
8 1 3 5 3 2 0 5 9 3 6 2 2 0 1 8 . 7
2 7 4 9 0 7 2 3 . 8 7 1 4 0 . 0 7 8
6 2 2 0 3 5 3 0 2 9 1 2 . 5 4 2 6 .
5 3 0 8 9 3 . 4 3 7 5 2 6 3 7 4 9 7
4 3 5 4 7 5 7 1 6 . 9 2 8 4 2 0 0 1
0 5 9 3 . 6 0 5 4 0 4 3 6 . 7 3 . 4
. 4 4 2 8 1 7 3 6 0 4 3 6 2 9 5 3 7
0 2 8 6 7 2 1 2 0 6 3 4 7 . 6 3 6 .
2 1 . . 4 . 9 6 5 . 0 1 2 9 9 1 4 0
8 6 4 8 2 1 5 7 4 6 2 7 1 3 5 . 2 1
. 3 0 9 9 . 0 9 6 7 9 8 8 . 7 6 2 7

Percentages Quiz (page 27)

1. 40%
2. 67%
3. 89%
4. 232%
5. 34.29%
6. 9%
7. 303%
8. 654%
9. 0.8%
10. 0.15
11. 0.004
12. 5.12
13. 0.0031
14. 17.65
15. 0.02
16. 1.00
17. 0.00287
18. 0.78
19. 119.25
20. 5
21. 107.8125
22. 771.75
23. 875
24. 410

Percentages Hidden Picture Puzzle (page 28)

1. 156.4
2. 4.5
3. 0.8
4. 26.25
5. 8.36
6. 16.25
7. 0.098
8. 361
9. 144
10. 12.15
11. 0.25
12. 10.62
13. 8.514
14. 171.92
15. 16.038
16. 1.2
17. 1.275
18. 163.02
19. 0.0665
20. 1.98

Percentages Hidden Picture Puzzle (page 28) cont.

79.8	179	378	5.672	45	983	2074	211	90	6501	63.22
425	1.275	0.25	1.06	12.98	616	7053	0.418	1.98	0.8	107
35	16.25	8.36	512	7.32	24.09	532	906	144	171.92	92
135	17.62	7.54	139	90.3	267	5223	431	1.95	82.08	11.567
569	293	3602	4576	163.02	10.62	26.25	1.86	3.554	6.22	6.8
12.34	61.52	487	6.35	156.4	361	16.038	728	1.342	39.5	34
46	82.67	287.3	14.12	790	18.63	6054	1.683	952	10.87	105.1
34.90	726	147	8.514	5.832	67.20	2.39	12.15	10.15	521	82.6
799	1462	555	14.09	0.0665	1.2	4.5	195.2	898	633.2	9653
385	4.754	78.2	643	153.9	0.098	76.1	383.6	62	14.13	80.2
16548	909	7.665	1.8325	7.85	58	311	6.9	0.364	28.9	0.4

Percentages Crossword Puzzle (page 29)

Percentages Connect-the-Dots Puzzle (page 30)

1. fraction
2. $\frac{73}{100}$
3. 73
4. numerator
5. denominator
6. point
7. two
8. right
9. percent
10. decimal
11. a) $83.30
    b) $14.70
    c) yes
12. a) $22.10
    b) $11.90
    c) $56.10

The connect-the-dots picture should be roughly in the shape of a percent sign.

Angles Word Search (page 32)

1. acute
2. right
3. obtuse
4. supplementary
5. complementary
6. corresponding
7. vertical
8. alternate interior
9. adjacent
10. alternate exterior

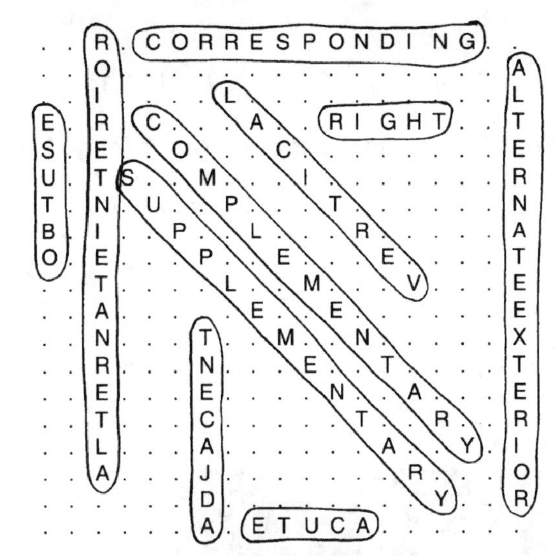

Geometric Figures Quiz (page 33)

Pentagon: 5, 2, 3, 180°, 540°, 108°
Hexagon: 6, 3, 4, 180°, 720°, 120°
Heptagon: 7, 4, 5, 180°, 900°, 128.57°
Octagon: 8, 5, 6, 180°, 1080°, 135°
Decagon: 10, 7, 8, 180°, 1440°, 144°
Dodecagon: 12, 9, 10, 180°, 1800°, 150°

Geometric Figures Crossword Puzzle (page 34)

Algebra: Algebraic Equations Quiz (page 35)

	Coefficient(s)	Variable(s)	Constant(s)
1.	7	x	3
2.	$\frac{1}{4}$	n	-9
3.	12, 6, -4	a, b, c	2
4.	3	y	11
5.	$\frac{2}{5}$, 8	g, m	-10

6. 35 7. 4 8. -19 9. 29 10. 15

Algebraic Equations Crossword Puzzle (page 36)

ACROSS	DOWN
1. $\frac{25}{x} + 3 = 8; x = 5$	1. $\frac{75}{x} = 5; x = 15$
2. $9 + \frac{20}{x} = 11; x = 10$	3. $6x - 3 = 45; x = 8$
4. $3x + 2 = 11; x = 3$	5. $5x - 10 = 50; x = 12$
6. $-6 + \frac{x}{2} = 5; x = 22$	7. $7 + 4x = 51; x = 11$
9. $9x - 14 = 67; x = 9$	8. $2x + 7 = 67; x = 30$
11. $18 - 2x = 14; x = 2$	10. $9 + 7x = 37; x = 4$
12. $\frac{21}{x} + 10 = 13; x = 7$	12. $\frac{60}{x} + 5 = 15; x = 6$
13. $x + 12 = 28; x = 16$	

Algebra: Rectangular Coordinate Systems Quiz (page 37)

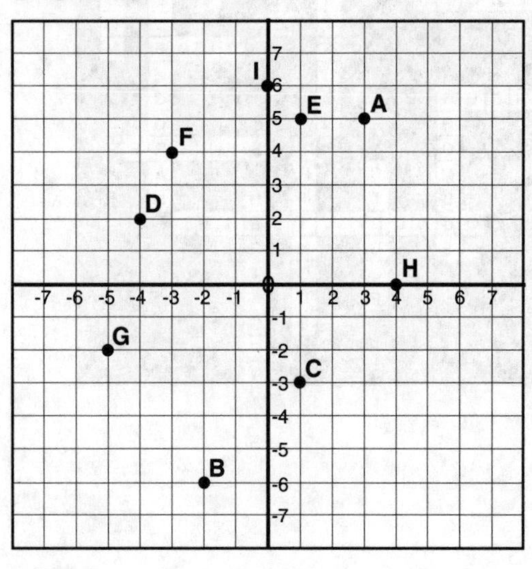

Rectangular Coordinate Systems Puzzle (page 38)

A.	(0, 12)	F.	(0, 0)
B.	(3, 7)	G.	(-8, -9)
C.	(10, 7)	H.	(-5, 1)
D.	(5, 1)	I.	(-10, 7)
E.	(8, -9)	J.	(-3, 7)

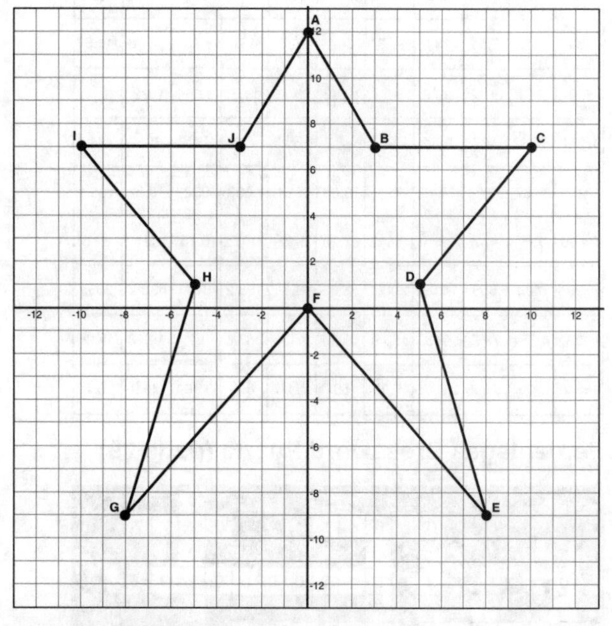

Metrics Quiz (page 39)

1. b
2. c
3. b
4. a
5. c
6. a
7. b
8. a

Metrics Crossword Puzzle (page 40)

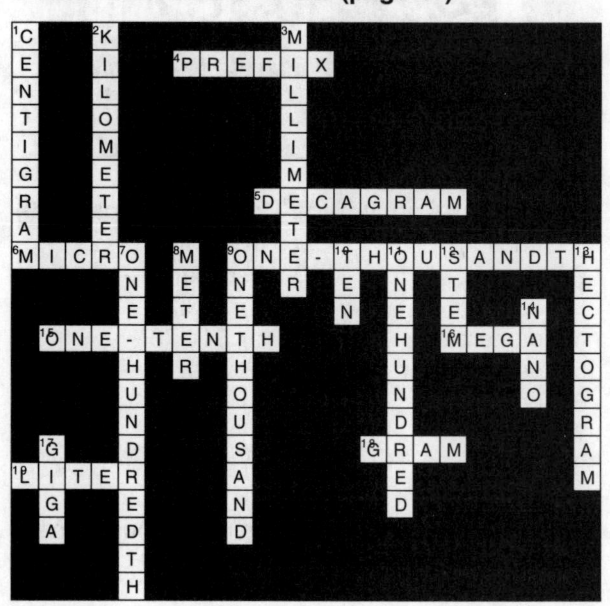